THE COMSOC GUIDE TO
NEXT GENERATION
OPTICAL TRANSPORT

THE COMSOC GUIDE TO NEXT GENERATION OPTICAL TRANSPORT
SDH/SONET/OTN

HUUB VAN HELVOORT, M.S.E.E.
Networking Consultant, The Netherlands
Senior Member IEEE

The ComSoc Guides to Communications Technologies
Nim K. Cheung, Series Editor

Celebrating 125 Years
of Engineering the Future

A John Wiley & Sons, Inc., Publication

Published by John Wiley & Sons, Inc., Hoboken, New Jersey.
Published simultaneously in Canada.

For general information on our other products and services or for technical support, please contact our Customer Care Department within the United States at (800) 762-2974, outside the United States at (317) 572-3993 or fax (317) 572-4002.

Wiley also publishers its books in a variety of electronic formats. Some content that appears in print may not be available in electronic formats. For more information about Wiley products, visit our web site at www.wiley.com.

Library of Congress Cataloging-in-Publication Data is available.

ISBN: 978-0-470-22610-0

To my ophthalmologist, who enabled me to see clearly what I wrote.

CONTENTS

PREFACE

Many times I have been asked to explain "*briefly*" how SDH, SONET, and the OTN "*exactly*" work. The questions came mainly from new colleagues, students, and users of these technologies, personally or via the usenet newsgroup comp.dcom.sdh-sonet. I could have referred them to the standards documents, but to provide a more consistent and clear answer I decided to write this pocket guide. The objective of this book is that it can be used both as an introduction as well as a reference guide to these technologies and their specific standards documents.

SDH (*Synchronous Digital Hierarchy*), SONET (*Synchronous Optical NETwork*), and OTN (*Optical Transport Network*) are transmission technologies mainly used in optical *Transport Networks* (TN). Digital information received from clients is transported in the payload area of containers (frame structures), which are synchronized to a common network clock. The containers also have an overhead area that contains information used for *Operation, Administration, and Maintenance* (OAM). The OAM is used to guarantee a high degree of *Quality of Service* (QoS) in a TN.

In the NGN (*Next Generation Network*), transport of packet-based client signals becomes more and more important. The extensions to SDH, SONET, and OTN to facilitate this transport are included in this book as well.

CHAPTER 1

INTRODUCTION TO OPTICAL TRANSPORT

This chapter covers the history of the development of SDH, SONET, and OTN. For consistency in the terminology used in this book I have included a list of conventions. This chapter ends with a list of all standards documents that are related to SDH, SONET, and OTN. Chapter 2 provides an overview of possible network topologies and introduces the functional modeling methodology used to describe the transport network. Chapter 3 describes the frame structures defined for use in the transport networks. In Chapter 4 the operation of SDH, SONET, and OTN is explained using the functional model. Per function it is described how the information in the associated frame structure is processed. The evolution of a pure voice-centered transport network towards a packet-based transport network (PTN) required the definition of a functional model of the latter as well. The model of the PTN is provided in Chapter 5. Processes that are generic and which can be used in different layers are described in separate chapters; the frequency justification processes in Chapter 6, the protection mechanisms in Chapter 7, and the mapping methodologies in Chapter 8. Chapter 9 describes how more flexibility in bandwidth can be achieved using existing frame structures. To conclude, Chapter 10 contains a reference table that provides an overview of the processing of the information in the SDH and SONET frame structures.

The ComSoc Guide to Next Generation Optical Transport: SDH/SONET/OTN,
by Huub van Helvoort
Copyright © 2009 Institute of Electrical and Electronics Engineers

1.1 HISTORY

Based on the experience with *Plesiochronous Digital Hierarchy* (PDH) equipment and networks, the telecommunication carriers and equipment vendors identified issues that should be resolved by developing a new methodology to transport digital signals. This resulted in a set of requirements:

- Standardize the optical interface specifications for the *Network Node Interface* (NNI) to enable the use of equipment from different vendors at both ends of a connecting optical fiber, i.e., the *"mid-span meet"* requirement.

  In PDH the signals at the distribution racks are specified but the line interfaces are proprietary requiring the installation of single vendor equipment at both ends.

- Use *"true"* synchronous multiplexing to simplify the access to any of the multiplexed signals in the network hierarchy. Synchronous multiplexing also simplifies the synchronization of the whole transport network to a single clock.

  In the PDH hierarchy it is required to demultiplex at every level to access the required signal. Each multiplex level has its own slightly different clock. Plesio-chronous = almost – synchronous

- Specify sufficient overhead capabilities to fully support automation of the operations, administration, and maintenance of the equipment and the transported signal.

  PDH has a very limited overhead capacity.

- Specify a multiplexing scheme that can be extended easily to accommodate higher bit-rates. It is expected to provide the transport infrastructure for worldwide telecommunications for at least the next two or three decades (starting in 1990).

  In PDH adding a higher multiplex is complex.

- The base rate of the new methodology should facilitate the transport of the highest PDH rate existing at that time.

The timeline for the development of the requirements mentioned above is:

1984   MCI raises the "mid-span meet" issue to various standards bodies: Exchange Carriers Standards Association (ECSA), Bell Communications Research (Bellcore, now Telcordia), American National Standards Institute (ANSI), and International Telecommunication Union (ITU-T, former CCITT).

1985   Bellcore proposes the concept for a *Synchronous Optical NETwork* (SONET). ANSI drives the standardization efforts (initially 400 proposals). The base rate is related to the DS-3 signal bit-rate of 45 Mbit/s. Frame structures are specified to transport DS-1, DS-2, and DS3.

1986   ITU-T initiates the development of *Synchronous Digital Hierarchy* (SDH) standards. The base rate is related to the E4 signal bit-rate of 140 Mbit/s. Frame structures are specified to transport E1, E2, E3, and E4. This is supported by the *European Telecommunications Standards Institute* (ETSI).

1987   An attempt to align SDH and SONET fails, the SONET base rate does not match any of the European PDH rates nor will it be able to transport E4 efficiently. There is also a significant difference in the frame format: SDH has 9 rows and 270 columns, while SONET has 13 rows and 180 columns.

1988   ANSI adjusts the SONET specification to meet the SDH requirements of the rest of the world. Agreement is reached on defining the base rate of SDH as 155 Mbit/s. The SONET STS-3 frame structure is the equivalent of the SDH STM-1.

1990   Multivendor SDH and SONET networks are operational.

See also the IEEE Communications Magazine, March 1989, "*SONET: Now It's the Standard Optical Network.*"

Since 1990 the SDH and SONET specifications have been extended based on demand for the transport of new tributary signals and also based on new capabilities provided by the evolution in component technology. The latter enabled the transport of multiple colors (wavelengths) on a single optical fiber. This was one of the reasons to start development of a new set of recommendations to use this feature: the *Optical Transport Network* (OTN), sometimes called "*digital wrapper.*" OTN has the capability to wrap any service into a digital optical container and thus enables network transparency that provides the flexibility to support all traffic types: voice, video, and data. The OTN technology seamlessly combines multiple networks and services into a common, future-ready infrastructure. Because it was designed to be transparent to the service type, all services carried over the OTN are given individual treatment, preserving any native functionality and performance without compromising the integrity of the underlying services.

2000   Specification of *Virtual conCATenation* (VCAT) to provide more flexibility in matching the bandwidth of the client signal. This is initiated by a high demand for transporting emerging technologies with non (SDH) standard bit-rates.

2001   Specification of the *Link Capacity Adjustment Scheme* (LCAS) to provide the ability to change the size of a VCAT signal. Specification of the *Generic Framing Procedure* (GFP) used for mapping packet-based signals into the constant bit-rate SDH signals. These two methodologies enable the use of SDH and SONET as transport capability in *Next Generation Networks* (NGN).

2001   The first version of the OTN recommendations is published.

2004   The specification of the Ethernet transport network is initiated by the ITU-T.

2005   The specification of a Transport Profile of MPLS (MPLS-TP) is started by the ITU-T and will be further developed in close cooperation with the IETF.

1.2 CONVENTIONS

In this book I have tried to cover both the SDH standards as specified, or recommended, by the global standardisation committee, the ITU-T, as well as the regional European standardization committee ETSI and the SONET standards specified by the regional standardization committee ANSI and the generic requirements specified by Telcordia. When appropriate I will mention differences between the SDH and SONET standards.

To avoid confusion that would be caused by mixing terminology and abbreviations used in the SDH, SONET, and OTN standards I will use a limited set of abbreviations and terms. However, when necessary, the equivalent SONET term is added in brackets {*term*}. To avoid even more confusion I will use abbreviations and terms that are already in use by the ITU-T.

1.2.1 SDH and SONET Terms

In this book I will use the term *Section* for the means for transportation of information between two network elements and make no distinction between an SDH Regenerator Section, in SONET termed {*Section*} and an SDH Multiplex Section, in SONET termed {*Line*}, i.e., the physical connection including the regenerators. Both SDH and SONET use the term *Path* for the connection through a network between the points where a container is assembled and dis-assembled. The total information transported over a path, i.e., the payload plus the OA&M information is in SDH normally referred to as a *Trail*.

- An SDH *Container* is the equivalent of a SONET *Synchronous Payload*. It is the general term used to refer to the payload area in a frame structure.
- An SDH *Virtual Container* (**VC**) is the equivalent of a SONET *Synchronous Payload Envelope* that exists in two forms, the {**STS-n SPE**} and the {**VTn SPE**}. It is used to refer to a **Container** and its associated **OverHead (OH)** information fields.
- **C–n**—a continuous payload *Container* of order n (n = 4, 3, 2, 12, 11). Generally used for all multiplex levels present in the STN. Sometimes used only to indicate the higher order levels of multiplexing (n = 4, 3).

- **C–m**—a continuous payload *Container* of order m (m = 2, 12, 11). Commonly used for the lower order levels of multiplexing.

- **C–n–X**—a contiguous concatenated payload *Container* of size X times the size of a container C–n. For (n = 4, 3) the X = 1...256 and for (n = 2, 12, 11) the X = 1...64.

- **VC–n**—a *Virtual Container* of order n (n = 4, 3, 2, 12, 11), that transports a container C–n together with its associated VC Path OverHead (VC POH). A VC–4 can also be sub-structured to transport three TUG–3s and similarly a VC–3 can transport seven TUG–2s. The SONET equivalents for the higher order multiplexes are for the VC–4: the *Synchronous Transport Signal* {STS–3c SPE}, and for the VC–3: the {STS–1 SPE}.

- **VC–m**—a *Virtual Container* of order m (m = 2, 12, 11), that transports a container C–m together with its associated VC POH. Commonly used for the lower order levels of multiplexing. The SONET equivalents for the lower order multiplexes are for the VC–2: the *Virtual Tributary* {VT6 SPE}, for the VC–12: the {VT2 SPE}, and for the VC–11: the {VT1.5 SPE}.

- **VC–n–Xc**—a *Contiguous concatenated VC-n* of order n (n = 4, 3, 2), that transports a container C–n–X together with its associated VC POH. The SONET equivalents are for the VC–4–Xc: the {STS–3–Xc} and for the VC–3–Xc: the {STS–1–Xc}.

- **VC–n–Xv**—a *Virtual concatenated VC-n* of order n (n = 4, 3, 2, 12, 11), that transports a container C–n–X by using X individual VC–n and their associated VC POH. The SONET equivalents are the {STS–3c–Xv SPE}, the {STS–1–Xv SPE} and the {VTn–Xv SPE}.

- **TU–n**—a *Tributary Unit* of order n (n = 2, 12, 11), that transports a VC–n together with its associated VC pointer.

- **TUG–n**—a *TU Group* of order n (n = 3, 2), that provides the flexibility for mapping tributaries. A TUG–3 can accommodate a single TU–3 or seven TUG–2s. A TUG–2 can accommodate a single TU–2, three TU–12s, or four TU–11s. The SONET equivalent is a {VT Group}.

- **AU–n**—an *Administrative Unit* of order n (n = 3, 4, 4-Nc and N = 4, 16, 64, 256), that transports a VC–n together with its associated VC pointer.

- **AUG–N**—an *AU Group* of multiplex level N (N = 1, 4, 16, 64, 256), that provides the multiplexing capability. For N = 4, ..., 256 it either accommodates a single AU–4–Nc or four AUG–N', where N' = N/4. An AUG–1 can accommodate a single AU–4 or three AU–3.

- **STM-N**—a *Synchronous Transport Module* of multiplex level N (N = 1, 4, 16, 64, 256), that transports an AUG-N together with the Multiplex Section overhead (MS OH) and the Regenerator Section overhead (RS OH). Equivalent SONET structures are the {STS-M where M = 3, 12, 48, 192, 768}.

1.2.2   OTN Terms

- **OPUk**—an *Optical channel Payload Unit* of order k (k = 1, 2, 3), the equivalent of an SDH C–n. It has its own payload-associated overhead.
- **OPUk–Xv**—a *Virtual concatenated OPUk*. Each of the X OPUk in an OPUk–Xv is transported as an individual OPUk in the OTN.
- **ODUk**—an *Optical channel Data Unit* of order k (k = 1, 2, 3), the equivalent of an SDH VC–n, that transports an OPUk. It has its own path-associated overhead.
- **ODUk–Xv**—a *Virtual concatenated ODUk*. Used to transport an OPUk–Xv.
- **OTUk**—an *Optical channel Transport Unit* of order k (k = 1, 2, 3), the equivalent of an SDH TUG–n. It has its own transport-associated overhead.
- **OTM–n,m**—an *Optical Transport Module*, n represents the maximum number of supported wavelengths and m represents the (set of) supported bit-rate. Similar to the SDH STM-N it has its associated Optical Multiplex Section (**OMS**) overhead and Optical Transmission Section (**OTS**) overhead.

1.2.3   Drawing Conventions

The order of transmission of information in all the figures in this book is first from left to right, and then from top to bottom. In frame structures the information in the first row is transmitted first, followed by the information in the second row, etc. Within each byte or octet the most significant bit is transmitted first. The Most Significant Bit (MSB) (bit 1) is shown at the left side in all the figures and the Least Significant Bit (LSB) at the right side.

1.3   STANDARD DOCUMENTS FROM DIFFERENT SDOS

Table 1-1 lists all the known standards and recommendation specifically related to SDH, SONET, and OTN developed by the international standardization organization and two regional standards organizations. A subdivision is made to group the specifications of different subjects used in the implementation of equipment and deployment in the transport network.

TABLE 1-1. List of standards

	ITU-T Recommendation[1]	ETSI Standards[1]	ATIS/ANSI Standards[2]
Source:	www.itu.int	www.etsi.org	www.atis.org
Physical interfaces	G.703 G.957 G.959.1 (OTN) G.691 G.692	EN 300 166 EN 300 232	T1.102 T1.105.06 T1.416 T1.416.01 T1.416.02 T1.416.03 *(GR 1374)*
UNI/NNI interfaces	G.8012 (ETH) G.8112 (MPLS-TP)	—	—
Network architecture	G.803 G.805 G.809 G.8010 (ETH) G.8110 (MPLS-TP)	ETR 114	T1.105.04
Structures & mappings	G.704 G.707 G.709 (OTN) G.7041 (GFP) G.7042 (LCAS) G.7043 (PDH-VCAT) G.8040 (GFP/PDH)	EN 300 167 EN 300 147 ETS 300 337	T1.105 T1.105.02 *(GR 253)*
Equipment functional characteristics (models)	G.783 G.798 (OTN) G.8021 (ETH) G.8121 (MPLS-TP) G.781 (sync) G.806 (generic)	EN 300 417-1-1 EN 300 417-2-1 EN 300 417-3-1 EN 300 417-4-1 EN 300 417-5-1 EN 300 417-6-1 EN 300 417-7-1 EN 300 417-9-1 EN 300 417-10-1 ETS 300 635 ETS 300 785	*(GR 496)* *(GR 499)* *(GR 2979)* *(GR 2996)*
Laser safety	G.664	—	—
Transmission protection	G.841 G.842 G.8031 (ETH linear) G.8032 (ETH ring) G.8131 (MPLS-TP lin.) G.8132 (MPLS-TP ring) G.808.1 (generic)	ETS 300 746 EN 300 417-1-1 EN 300 417-3-1 EN 300 417-4-1 TS 101 009 TS 101 010	T1.105.01 *(GR 1230)* *(GR 1400)*

TABLE 1-1. *Continued*

	ITU-T Recommendation[1]	ETSI Standards[1]	ATIS/ANSI Standards[2]
Restoration	M.2102	—	
Information model	G.774	ETS 300 304	T1.119
	G.774.01	ETS 300 484	T1.119.01
	G.774.02	ETS 300 413	T1.119.02
	G.774.03	ETS 300 411	T1.245
	G.774.04	ETS 300 493	*(GR 836)*
	G.774.05	EN 301 155	*(GR 1042)*
	G.774.06		*(GR 2950)*
	G.774.07		
	G.774.08		
	G.774.09		
	G.774.10		
	M.3100 (generic)		
Equipment management	G.784	EN 301 167	*(GR 3000)*
	G.8051 (ETH)	EN 300 417-7-1	
	G.8151 (MPLS-TP)		
Network management	G.831	ETS 300 810	T1.204
	G.850–G.859		*(GR 3001)*
Management communications interfaces	G.773		T1.105.04 *(GR 376)*
Error performance [equipment level]	G.783	EN 300 417-1-1	—
	G.784	EN 300 417-2-1	
	O.150	EN 300 417-3-1	
	O.181	EN 300 417-4-1	
	O.182 (OTN)	EN 300 417-5-1	
		EN 300 417-6-1	
		EN 300 417-7-1	
		EN 300 417-9-1	
		EN 300 417-10-1	
Error performance [network level]	G.826	EN 301 167	T1.105.05
	G.827		T1.231
	G.828		T1.514
	G.829		*(GR 2991)*
	G.8201 (OTN)		
	M.2101		
	M.2101.1		
	M.2102		
	M.2110		
	M.2120		
	M.2130		
	M.2140		

TABLE 1-1. *Continued*

	ITU-T Recommendation[1]	ETSI Standards[1]	ATIS/ANSI Standards[2]
Jitter & wander performance	G.813 G.822 G.823 G.824 G.825 G.8251 (OTN) G.783 O.172 O.173 (OTN)	EN 300 462-5-1 EN 302 084	T1.105.03 T1.105.03a T1.105.03b
Synchronization [clocks & network architecture]	G.803 G.810 G.811 G.812 G.813 G.8261 (PTN) G.8252 (ETH)	EN 300 462-1 EN 300 462-2 EN 300 462-3 EN 300 462-4 EN 300 462-5 EN 300 462-6 EN 300 417-6-1 EG 201 793	T1.101 T1.105.09

Note 1: The ITU-T recommendations and ETSI standards can be downloaded freely from www.itu-t.int and www.etsi.org, respectively.

Note 2: The ANSI standards can be ordered at www.atis.org. This column also lists Telcordia documents (*in italics*), which can be ordered at www.telcordia.com.

CHAPTER 2

NETWORK ARCHITECTURES: THE STRUCTURE OF THE NETWORK

Initially digital multiplexers were used in point-to-point connections between telephony switching equipment. Later, together with the evolution of the telephone network, higher levels of multiplexing were introduced to connect local exchanges (central office) and long distance exchanges (toll office). The hierarchy of the telephony network determined the structure of the first transport network topology: the PDH hierarchy. Based on the experience gained with this initial topology, the topology for the SDH and SONET networks was designed. The first part of this chapter describes the different topologies.

One of the objectives for the design of the SDH technology was to improve the operation, administration, and management of the network and the equipment in the network nodes. After a thorough study the functional modeling methodology was developed to provide a means to specify and describe the transport network in a uniform and unambiguous way. The layered model was introduced, where each layer can be operated and managed independent of other transport layers. The network layering principles and the atomic functions that are present in each layer are described in the second part of this chapter.

2.1 NETWORK TOPOLOGY

The structure of SDH, SONET, and OTN networks is based on the experience gained with PDH networks. Several topologies can be recognized.

The ComSoc Guide to Next Generation Optical Transport: SDH/SONET/OTN,
by Huub van Helvoort
Copyright © 2009 Institute of Electrical and Electronics Engineers

2.1.1   PDH Hierarchy

The structure of PDH networks was very hierarchical; it was similar to the hierarchical structure of the telephony network. Several levels can be recognized based on the multiplex rate: primary rate, secondary rate, ternary rate, etc. An example of such a hierarchy is drawn in Figure 2-1. The multiplexers at each level in the network have the capability to cross-connect and multiplex lower rate signals. The multiplexed signal can be sent to multiplexers at the same level or to higher level multiplexers for further transport in the network.

2.1.2   Linear Topology

For point-to-point connections in a simple network two *Terminating Multiplexers* (TM) are required. A TM has two sides, a customer or tributary side that supports one or more tributary signals, and a network or line side that supports a line signal. A TM multiplexes tributary signals, e.g., PDH signals, into a line signal, e.g., an SDH STM-1, SONET OC-48, or OTN OTM-0. It is possible to insert between the two TMs one or more *Add Drop Multiplexers* (ADM). An ADM has three sides, one tributary side and two line sides also referred to as east side and west side. An ADM is capable of multiplexing tributary signals for insertion (Add) into one of the two line interfaces, or demultiplexing a tributary signals (Drop) from one of the two line interfaces, and also passing through tributary signals from one line interface to the other. Figure 2-2 shows an example of a linear topology.

To improve the reliability of the line signals, they can be duplicated and a bridge and selector can be used to select the best signal. This protection switch-

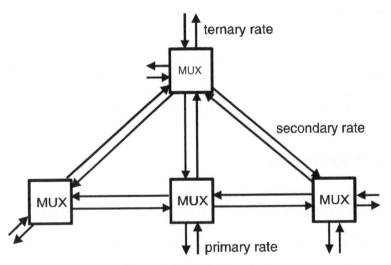

Figure 2-1. PDH hierarchy

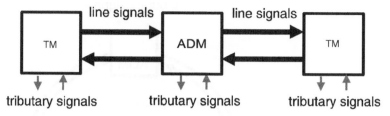

Figure 2-2. Linear topology

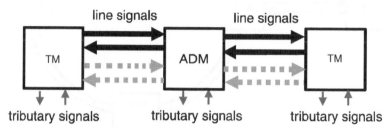

Figure 2-3. Linear topology—protected

ing is described in Chapter 7. The duplication is shown in Figure 2-3 by the dashed gray lines. Note that these additional or protecting signals have to be transported over a physical path that is separated from the original signals to avoid single points of failure.

Figure 2-3 also shows that the ADM in the middle has become a single point of failure in this topology, especially for the signals that it passes through from one TM to the other.

To resolve this, either the ADM can be replaced by two ADMs, which should be physically separated, or the additional signals can bypass the ADM. The result will be a different topology: the ring, which is described next.

2.1.3 Ring Topology

A ring topology consists of a number of ADMs, i.e., the ring nodes, where each ADM is connected to two other ADMs by an optical fiber pair, the ring section or span, which transports the line signals. This is depicted in Figure 2-4.

Each ADM in the ring can be reached via two different paths. The ring topology provides a high reliability, especially if protection mechanisms, which are described in Chapter 7, are applied.

Tributary signals can be connected to the ADMs and transported over the ring. It is also possible to connect a TM to the tributary side of an ADM if the ADM supports tributary SDH/SONET/OTN signals. These tributaries are generally one or more multiplexing levels lower than the line signals of the

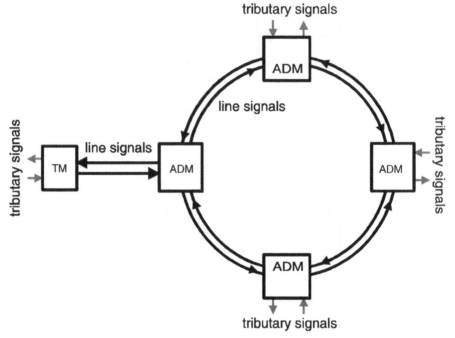

Figure 2-4. Ring topology

ADM. For example, in an SDH network an ADM with STM-16 (2.4 Gbit/s) line signals may have the capability to support STM-1 (155 Mbit/s) and/or STM-4 (633 Mbit/s) tributary interfaces. **Note:** *the STM-N frame structure is defined in Section 3.1.1 of Chapter 3.* To improve the reliability of the connection between the TM and the ADM it is connected to, again a ring can be created using a lower line signal rate. This is shown in Figure 2-5.

In this example the STM-4 ring can be used as an aggregation ring that collects tributary traffic and the STM-16 ring can be used as a transport or back-bone ring to transport traffic between the aggregation rings.

Again the ADM that connects the two rings is a potential single point of failure, especially for signals that are passed through from one ring to the other. To resolve this, the ADM can again be replaced by two ADMs such that the two rings are interconnected in two nodes. If this is applied in several places in the network, the ring topology will evolve into a mesh topology, as described next.

2.1.4   Mesh Topology

A mesh topology consists of *Digital Cross Connect* (DXC) systems, but also ADMs and TMs may be present. A DXC is a system that has multiple line

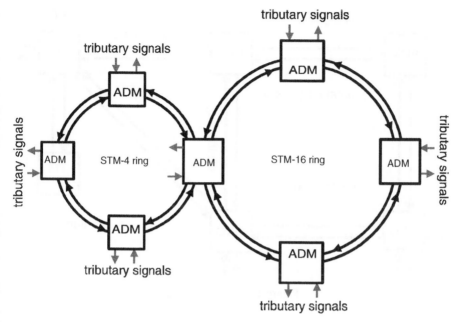

Figure 2-5. Dual ring topology

interfaces and a cross-connect function to switch the transported tributary signals between any of the line interfaces. A DXC may have tributary interfaces for local access. Figure 2-6 shows a small example of a mesh or maze topology.

A mesh topology can be considered as a set of interconnected rings. However, contrary to the ring topology, the data rate of the interconnecting line signals is not necessarily the same on all spans.

2.1.5   NGN Topology

For a long time the transport network was primarily used for the transport of voice signals. Initially the analogue voice signals were multiplexed using analogue *Frequency Division Multiplex* (FDM) for the transport between telephone exchanges. After the invention of *Pulse Coded Modulation* (PCM) in 1935 and the invention of the transistor in 1948 the FDM systems were replaced by digital Time Division Multiplex (TDM) systems and the PDH network was formed.

For the communication between computers, initially modems were used. The modem converted the digital data stream into an analogue signal for the transport over the telephone line. Later the modems were replaced by proprietary PDH links.

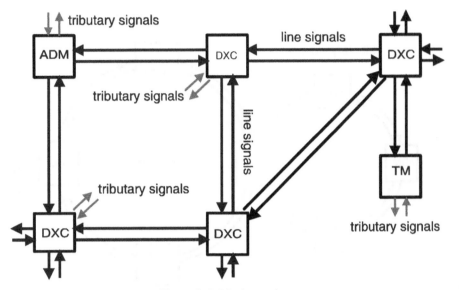

Figure 2-6. Mesh topology

Also, the SDH technology was initially designed to transport TDM, i.e., PDH signals.

The real breakthrough came around the year 2000 when the demand for data transport grew explosively. After the "internet hype" was over it became clear that the network bandwidth required for the transport of data would soon surpass the bandwidth required for voice transport. Data that is transported does not have constant bit-rate because it is transferred in packets that may vary in length and will be sent only when data is present.

Originally the transported digital data was restricted to information exchanged between computers. Later on, as a result of progress in technology, voice and video signals are also transported using the packet technology. The transport of these different services is referred to as triple-play: voice, video, and data are all transported in packets.

Because the SDH and OTN were initially designed to transfer voice signals, they needed to be extended to provide the capability to support packet based technologies. The flexibility to provide the bandwidths matching the data rates is achieved by the definition of *Virtual Concatenation* (VCAT) as described in Chapter 9. The mapping of the packets into the SDH and OTN containers is achieved by the definition of the *Generic Framing Procedure* (GFP) as described in Chapter 8.

This is considered the start of the *Next Generation Network* (NGN). In the NGN the basic network elements, like ADM ad DXC, are replaced by *Multi-Service Transport Platforms* (MSTP) that are capable of transporting voice and data signals over the SDH/SONET/OTN network.

In the transport network several types of MSTP can be distinguished:

- *Multi-Service Access Platforms* (MSAP)—platforms located at the edge of the network. They are used to collect the data from the individual customers. At the tributary side they support TDM signals, e.g., PDH and lowest order SDH/SONET, and packet-based signals, e.g., *Fast-Ethernet* (FE). The line interfaces are STM-1 {OC-3} and STM-4 {OC-12}.
- *Multi-Service Provisioning Platforms* (MSPP)—platforms located in the metro area or edge aggregation area of the network. They are used to aggregate the information collected in the access rings (or by a *Passive Optical Network* (PON)). At the tributary side they support STM-1 {OC-3} and STM-4 {OC-12} signals as well as *1 Gigabit Ethernet* (1GbE) and *Fibre Channel* (FC) signals. The line interfaces are STM-4 {OC-12} and STM-16 {OC-48}.
- *Multi-Service Switching Platforms* (MSSP)—platforms located in the high-capacity core of the network. At the tributary side they support STM-16 {OC-48} as well as high-speed data rates, e.g., *10 Gigabit Ethernet* (10GbE). At the line side they support interfaces such as STM-64 {OC-192} or STM-256 {OC-768}, but also OTU2 or OTU3 may be supported in OTN networks.

Figure 2-7 shows the possible location of these platforms in an example of a transport network. The platforms allow the service providers to build an NGN that integrates core and edge networks and can handle voice, video, data, and other services.

They provide the traffic management and connectivity capabilities needed to implement *Virtual Private Networks* (VPN). With these platforms the service

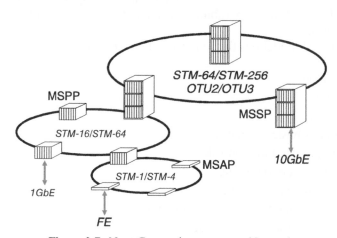

Figure 2-7. Next Generation transport Network

providers can operate *Local Area Networks* (LAN), *Metro Area Networks* (MAN), *Wide Area Networks* (WAN), and *Storage Area Networks* (SAN) very efficiently. The added value is that the service providers can use their existing SDH, SONET, and OTN infrastructures, and also their legacy *Operations, Administration, and Management* (OAM) philosophy.

Note: *the figure shows only Ethernet tributaries (FE, 1GbE, 10GbE); other packet technologies are also possible with equivalent bit-rates.*

2.2 NETWORK FUNCTIONAL MODEL

The ITU-T uses the functional model methodology to describe equipment specifications for the different technologies. The same functional modeling can be used by service providers to specify their transport networks. Also the information model used to support the management of the network and the equipment is based on this methodology.

A functional model consists of atomic functions; each atomic function has capabilities that are equivalent in different technologies. In a transport network and in transport equipment layers can be identified that have the same characteristics. In each layer characteristic information is transported. Each layer can be partitioned to match the network or equipment topology. The functional modeling supports the layered model and partitioning to describe the network and the equipment.

The development of functional models for use in telecommunication networks was a combined effort of service providers and equipment makers. After an extensive analysis of existing transport network structures the functional modeling methodology was introduced first in the standards documents of ETSI around 1995, and the ITU-T adapted the functional modeling in 1997. Due to the increasing complexity and variety of the transport network and equipment it became almost impossible to guarantee compatibility and interoperability of equipment based only on written documentation.

A functional model has the following characteristics:

- it is simple,
- it is short,
- it is visual,
- it contains basic elements,
- it provides combination rules,
- it supports generic usage,
- it has recursive structures,
- it is implementation independent, and
- it is transport level independent.

2.2.1 Specification of the Atomic Functions

The atomic functions are specified such that they can be used to model a network or equipment and be able to use that model to provide a structured and well-organized set of requirements for the design and management of the actual network as accurately as possible. The atomic functions describe the processes that have to be performed to transport information between specific points in the model. The functional model allows partitioning a network and identifying layers that contain the same atomic functionality. This provides a high degree of recursion.

Atomic functions will comply with the following requirements:

- The functional model that uses the atomic functions shall describe the *functional behavior* of the implementation and *not* the implementation itself.
- The processes in the atomic functions shall *not* describe the underlying hardware and/or software architecture.
- The number of atomic functions shall be limited to keep the functional model simple.

The functional modeling methodology provides a *common language* that can be used at all levels involved in the deployment of a telecommunications network:

- specifying telecommunication standards recommendations;
- describing the transport network of a service provider;
- describing the equipment of a manufacturer;
- describing deliverables in contracts;
- describing the equipment architecture and requirements for the R&D department;
- describing device architecture and requirements for the AISC manufacturer.

2.2.3 Description of Atomic Functions

The result of the thorough analysis of the existing transport networks is a set of atomic functions that can be used to describe the functionality of a transport network in an abstract way by using only a small number of atomic functions. In general the atomic functions that are currently defined in the standards will process the information that is presented at one or more of their inputs and then present the processed information at one or more of their outputs. The atomic functions can be associated with each other following the connection rules to construct a network element. A transport network can be built using the models for the network elements. In the transport network architecture reference points can be identified that are the result of the binding of the inputs and outputs of processing functions and transport entities.

For each generic function a specific symbol has been defined as well as the connection rules. These functions and their symbols will be described next and are also illustrated in Figures 2-8, 2-9, and 2-10.

2.2.3.1 Connection Atomic Function

This function provides the connectivity in a layer present in a network element and in a network. It is identified by *<layer>*_C, where the *<layer>* shall be replaced by the identifier of the layer. Figure 2-8 shows the symbol used for a connection atomic function; it is normally drawn in bi-directional representation.

The connection function is defined for a connection-oriented network. The equivalent function in a connectionless network is referred to as a flow domain because the information is transferred in flows instead of over connections.

The bi-directional *Connection function <layer>*_C is the transport processing function that provides the connectivity in a layer network. The connections are made between *Connection Points* (CP), i.e., an input and one or more outputs. The information that passes the CP is characteristic for the layer and is referred to as *Characteristic Information* (CI). The CI of a particular layer is indicated by *<layer>*_CI. Sometimes the *Termination CP* (TCP) is used in the model to make a distinction between the CP of an adaptation atomic function and the CP of a trail termination atomic function, a TCP. In general bi-directional connections are provisioned between a TCP and a CP. However, bi-directional connections between a TCP and a TCP, or between a CP and a CP are also possible. Connections can be provisioned by the *Element Management Function* (EMF).

The Connection function *inputs and outputs:*

- The *<layer>*_C function has one or more input (T)CP and one or more output (T)CP. The following processes can be present in a connection function either as a single process or in combination with others:
  - Routing and grooming.
  - Protection switching based on server layer status: *Server Signal Fail* (SSF) and *Server Signal Degrade* (SSD).

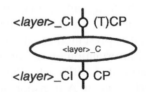

Figure 2-8. Connection atomic function

- Protection switching based on far-end status and operator request: *Automatic Protection Switching* (APS).
- Consequent actions, e.g., generation of *"idle"* or *"unequipped"* signals if an output is not connected to an input.

2.2.3.2 Adaptation Atomic Function This function provides the adaptation of the information structure present in the client layer network at its CP to the information structure present in the server layer network at its *Access Point* (AP).

Figure 2-9 shows the symbol of an adaptation atomic function, at the left side the unidirectional representation and at the right side the bi-directional symbol.

The adaptation function is identified by *<srvr>/<clnt>_*A, where *<srvr>* should be replaced by the identifier of the server layer and *<clnt>* shall be replaced by the identifier of the client layer. The information passing the AP is *Adapted Information* (AI) and is in general referred to as *<srvr>/<clnt>_*AI.

The uni-directional *Adaptation Source function <srvr>/<clnt>_*A_So is the transport processing function that adapts the client layer network characteristic information into a form suitable for transport over a trail in the server layer network.

The uni-directional *Adaptation Sink function <srvr>/<clnt>_*A_Sk is the transport processing function that recovers the characteristic information of the client layer network from the server layer network trail information.

The bi-directional *Adaptation function <srvr>/<cclnt>_*A is the transport processing function that consists of a co-located and associated Adaptation Source and Sink pair.

The Adaptation function *inputs and outputs:*

- The *<srvr>/<clnt>_*A_So function can have one or more CPs and has only a single AP. This configuration is commonly used to represent the multiplexing of several client signals into a single server signal. The client signals originate not necessarily from the same layer network.

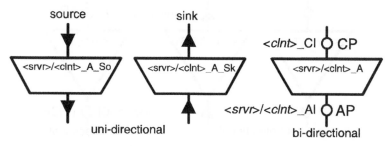

Figure 2-9. Adaptation atomic function

- The *<srvr>/<clnt>_*A_Sk function has one AP and can have several CPs. This configuration is used to represent inverse multiplexing. It is possible that the CPs are connected to different client layer networks.

The following processes can be present in an adaptation function either as a single process or in combination with others:

- Scrambling, coding—descrambling, decoding
- Rate adaptation, frequency justification
- Transfer protection switch status: *Automatic Protection Switching* (APS).
- Transfer server layer status to the connection function: *Server Signal Fail* (SSF) and *Server Signal Degrade* (SSD).
- Transfer synchronization status: *Synchronization Status Message* (SSM).
- Alignment
- Payload type identification: *Payload Signal Label* (PSL).
- Multiplexing and mapping
- Transfer communications channels: e.g., *Data Communications Channel* (DCC), *User Channel* (USR), and *Engineering Order Wire* (EOW).

2.2.3.3   Trail Termination Atomic Function   This function provides the start and end points, or *Source* and *Sink*, of a trail through the transport layer network. The trail starts at the AP and is connected to the connection function via the *Termination* CP (TCP). It is identified by *<srvr>_*TT, where *<srvr>* shall be replaced by the identifier of the server layer. Figure 2-10 shows the symbol of a trail termination atomic function, at the left side the unidirectional representation and at the right side the bi-directional symbol.

The *trail termination function* is defined for a connection-oriented network. The equivalent function in a connectionless network is referred to as a *flow termination function*.

The uni-directional *Trail Termination Source function <srvr>_*TT_So is the transport processing function that processes the adapted client layer charac-

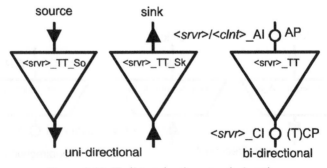

Figure 2-10. Trail termination atomic function

teristic information presented at its input and adds overhead information to provide monitoring of the trail in the layer network. The resulting characteristic information of the server layer network is presented at its output. A trail termination source function can operate without an input from a client layer network and used for testing purposes.

The uni-directional *Trail Termination Sink function* <srvr>_TT_Sk is the transport processing function that accepts the characteristic information of the server layer network at its input, removes the information related to the performance monitoring of the trail and presents the remaining client layer network information at its output. A trail termination sink function can operate without an output to a client layer network and used as trail monitoring function.

The bi-directional *Trail Termination function* <srvr>_TT is the transport processing function that consists of a co-located and associated Trail Terminations Source and Sink function pair.

The Trail Termination function *inputs and outputs:*

- The <srvr>_TT_So function always has one input AP. It generally has also a single output (T)CP. If the function has inverse multiplex capabilities, multiple outputs are present. An example is Virtual Concatenation, described in Chapter 9.
- The <srvr>_TT_Sk function can have one or more input (T)CP and always has a single output AP.

The following processes can be present in a trail termination function either as a single process or in combination with others:

- Scrambling–descrambling
- Error detecting code generation—checking: *Error Detection Code* (EDC).
- Trail identification—connectivity check: *Trail Trace Identifier* (TTI),
- Near-end and Far-end performance monitoring
- Status reporting at the near-end: *Trail Signal Fail* (TSF) and *Trail Signal Degrade* (TSD).
- Status reporting to the far-end: *Remote Defect Indication* (RDI) and *Remote Error Indication* (REI),

2.2.4   Atomic Functions Reference Points

The CP and AP are not the only reference points that are defined in the functional model. The following reference points can also be present, but may be shown only when their presence in the model is required.

- *Management Point* (MP)—provides the input and output for management information, e.g., provisioning, fault reporting, performance monitoring.

- *Timing Point* (TP)—provides the input and output for timing and synchronization information, e.g., clock and framestart signals.
- *Remote Point* (RP)—provides the connection between co-located and associated sink and source functions to pass information that has to be sent to the far-end, e.g., RDI and REI.
- *Replication Point* (PP)—provides the connection between transport Ethernet adaptation source and sink functions to transfer replicated information.

2.2.5 Atomic Function Relations

The relation between the atomic functions and the relation to the layered network is illustrated in Figure 2-11. The figure also shows the order of successive atomic functions in a functional model. The order is recursive, so it is the same in each layer. When the connection atomic function consists of a single connection and contains no processing it can be omitted from the model. The right side in the figure shows the names of the functions as they are used in a connection-oriented technology, e.g., SDH. The left side shows the names used in connectionless oriented technologies, e.g., transport Ethernet. The description of the connection-oriented atomic functions is based on [ITU-T Rec. G.805]. The description of the connectionless atomic functions can be found in [ITU-T Rec. G.809].

A more extensive description of functional modeling can be found in the book "*SDH/SONET Explained in Functional Models: Modeling the Optical Transport Network*", John Wiley & Sons, ISBN 0-470-09123-1.

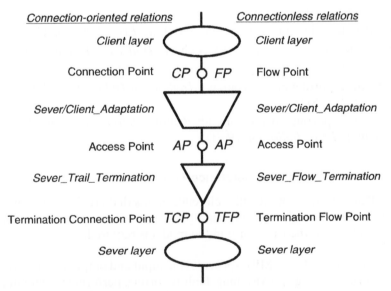

Figure 2-11. Functional model logical order

2.2.6   Atomic Functions and Transport Entities

In a functional model the transport entities take care of the transparent transfer of information between two or more layer network reference points. The information presented at the output is exactly the same information received at the input unless it is affected by degradation, e.g., bit errors, of the transfer process.

Two basic transport entities are distinguished:

- *Trails*, this is the entity that represents the connection through the layer network of the information that has been adapted for the transport. It runs from the entry point to the exit point of a specific layer network.
- *Connections.* They can be subdivided into:
  - *Network connections* (NC), show how the information is transported in the specific layer network.
  - *Sub–network connections* (SNC), a representation of the switching matrix in the equipment for a specific layer network.
  - *Link connections* (LC), represents the section of a network connection that is used to transport the information in a server layer.

Figure 2-12 shows the transport entities described above.

Note: *The concept of client layer networks and server layer networks in the transport network model is completely orthogonal to the layering concept used*

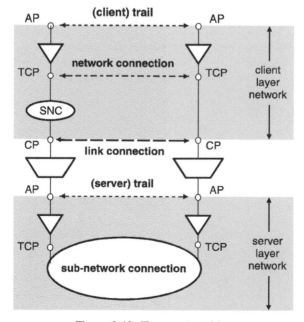

Figure 2-12. Transport entities

in the OSI protocol reference model described in [ITU-T Rec. X.200]. *All the 7 OSI layers are present in each of the transport network layers, some of which may be void in a particular layer:*

- *The* adaptation *function provides the* application *(7) and* presentation *(6) layer,*
- *The* termination *function provides the* session *(5),* network *(3),* data link *(2), and* physical *(1) layers,*
- *The* connection *function provides the transport (4) layer.*

CHAPTER 3

FRAMES AND STRUCTURES: TRANSPORT IN CONTAINERS

In the *Synchronous Transport Network* (STN, a general term for SDH and SONET) and the *Optical Transport Network* (OTN) the signal structures are described using frames that have two dimensions: rows and columns. Generally the number of rows is fixed; there are nine rows in SDH/SONET structures and four rows in OTN structures.

To provide flexibility in the definition of signals with different bit-rates in SDH and SONET the number of columns is variable while the frame repetition rate is fixed at 125 μsec, i.e., the 8 kHz clock that is used as a reference in the whole network. In the OTN the number of columns is fixed at 3824 while the frame repetition rate is variable. The structure of PDH signals is also frame based, but in general one dimensional. In the *Packet Transport Network* (PTN) the signal structures, i.e., packets or cells, are described using frame structures that have a single dimension.

This chapter provides an overview of the STN, OTN, and PTN frame structures.

3.1 SDH FRAMES AND STRUCTURES

In SDH the following frames and structures can be distinguished: STM-N, AUG-N, AU-n, TUG-n, TU-n, VC-n, and C-n. The related standards are [ITU-

The ComSoc Guide to Next Generation Optical Transport: SDH/SONET/OTN,
by Huub van Helvoort
Copyright © 2009 Institute of Electrical and Electronics Engineers

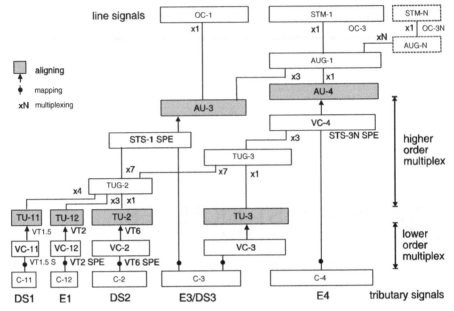

Figure 3-1. SDH and SONET multiplex structures

T Rec. G.707] and [ETSI EN 300 417-1 … 10]. SONET has similar frames and structures, i.e., OC-N, STS-1, VTn, STS-1 SPE, STS-3N SPE, and VTn SPE, which are specified in [ANSI T1.105].

Figure 3-1 shows the relation between these frame structures.

The alignment process is described in Chapter 6 and the mapping methodology in Chapter 8.

All these frame structures are described in the following sections.

3.1.1   STM-N Frame Structure

Existing SDH frame structures of the *Synchronous Transport Module of level N* (STM-N) are STM-0, STM-1, STM-4, STM-16, STM-64, and STM-256.

SONET makes a distinction between a physical signal, i.e., the *Optical Carrier of level N* (OC-N), and a logical signal, i.e., the *Synchronous Transport Signal of level N* (STS-N). The equivalent physical signals are OC-1, OC-3, OC-12, OC-48, OC-192, and OC-768, and the equivalent logical frame structures are STS-1, STS-3, STS-12, STS-48, STS-192, and STS-768.

The STM-N frame structure can be represented by a frame consisting of 9 rows and Y columns. For the values of N = 1, 4, 16, 64, and 256 the values of X and Y are: X = 9 × N and Y = 270 × N. For the value of N = 0, i.e., an {STS-1}, they are X = 3 and Y = 90.

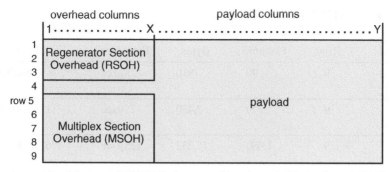

Figure 3-2. STM-N structure

The STM-N (N = 1, 4, 16, 64, 256) structure is divided into three specific areas as indicated in Figure 3-2 (the location in the frame is indicated by [*row, column*]):

- One area is allocated to the *Regenerator Section OverHead* (RSOH). It consists of bytes [1 ... 3, 1 ... (9 × N)] containing the OAM for the regenerators between two multiplexers. The use of OAM is described in Section 4.5 of Chapter 4: the RS layer.
- The second area is allocated to the *Multiplex Section OverHead* (MSOH). It consists of bytes [5 ... 9, 1 ... (9 × N)] containing the OAM for the two multiplexers. The use of OAM is described in Section 4.4 of Chapter 4: the MS layer.
- The third area is allocated to the payload that is transported. It consists of pointer bytes [4, 1 ... (9 × N)] and bytes [1 ... 9, (9 × N + 1) ... (270 × N)]. For the management system this is an *Administrative Unit Group of level N* (AUG-N), see also Section 3.1.2 below.

The STM-0 {STS-1} structure is divided as follows:

- The RSOH area consisting of bytes [1 ... 3, 1 ... 3].
- The MSOH area consisting of bytes [5 ... 9, 1 ... 3].
- The payload area consisting of bytes [4, 1 ... 3] and bytes [1 ... 9, 4 ... 90]. The payload area can fit exactly an *Administrative Unit of level 3* (AU-3), see section 3.1.3 below.

Multiplexing to a higher-level STM-N is achieved by column-interleave multiplexing four lower-level STM-N frame structures, except for the multiplex from STM-0 {OC-1/STS-1} to STM-1 {OC-3/STS-3c} that uses a factor three. Table 3-1 gives an overview of the possible line bit-rates.

TABLE 3-1. STM-N {STS-M} bit-rates

STM-N STS-M	Rows	Columns	Bytes	Frame-rate	Bit-rate
STM-0 STS-1	9	90	810	125 μsec	51.840 Mbit/s
STM-1 STS-3	9	270	2,430	125 μsec	155.520 Mbit/s
STM-4 STS-12	9	1,080	15,232	125 μsec	622.080 Mbit/s
STM-16 STS-48	9	4,320	51,840	125 μsec	2,488.320 Mbit/s
STM-64 STS-192	9	17,280	155,520	125 μsec	9,953.280 Mbit/s
STM-256 STS-768	9	69,120	622,080	125 μsec	39,813.120 Mbit/s

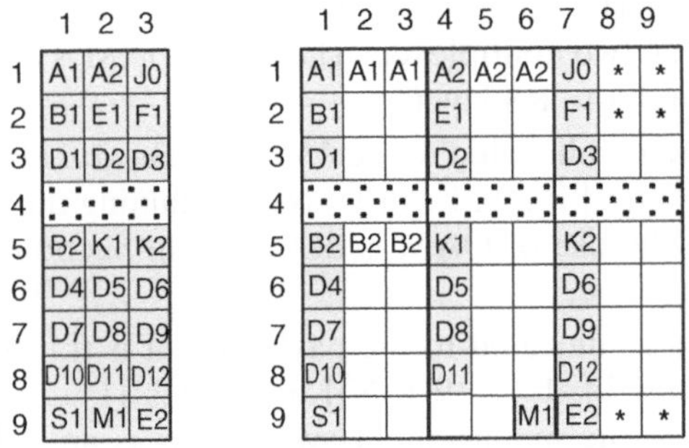

Figure 3-3. STM-0/STS-1 and STM-1/STS-3 overhead

Even though the overhead areas are also multiplexed, only the overhead bytes of the first STM-N in the multiplex are maintained. The overhead bytes of the successive STM-N are overwritten, except the A1, A2, and B2 bytes. The bytes identified with "*" are allocated for national use. Figure 3-3 shows the three times multiplex of an STM-0 to STM-1.

Note—*byte M1 moved from byte [9, 4] to byte [9, 6].*

Figure 3-4 shows the four times multiplex of an STM-1 to STM-4.

Figure 3-5 shows the generic STM-N frame structure with the basic allocated overhead bytes: A1 and A2 for frame alignment, J0 for identification, B1 and B2 for performance monitoring, E1 and E2 are engineer order wires, F1 is a user channel, D1 … D3 and D4 … D12 provide data communication

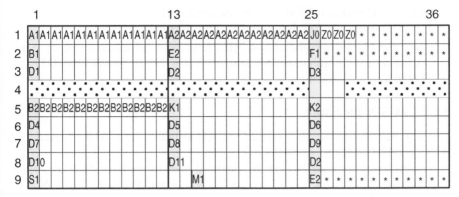

Figure 3-4. STM-4/STS-12 overhead

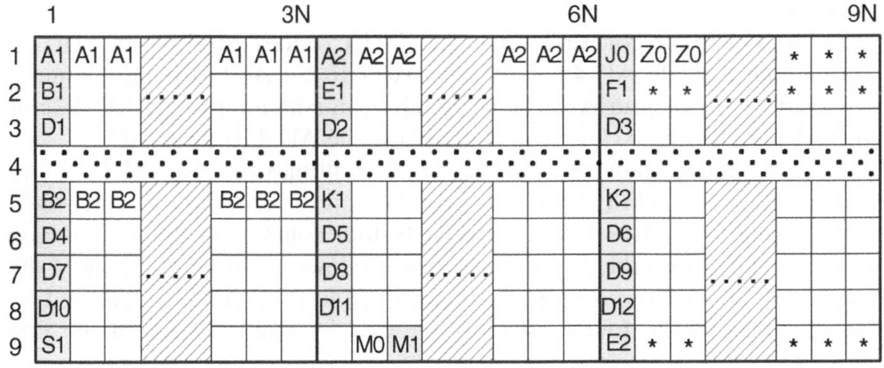

Figure 3-5. STM-N/STS-3N overhead

channels, K1 and K2 for protection switching control and remote defect indication, S1 for synchronization status, and M1 for remote error indication.

In the higher order multiplexes additional bytes have been designated to a specific use. The following bytes are used:

- In the STM-4, STM-16, STM-64, and STM-256 overhead (N-1) bytes are reserved for future international standardization, these Z0 bytes are located in $[1, (6N + 2) \ldots 7N]$.
- In the STM-16, STM-64 and STM-256 overhead a number of bytes are allocated to contain a *Forward Error Correcting* (FEC) code. One byte, i.e., Q1, is the FEC control byte, and the P1 bytes contain the FEC code. An STM-16 has 3×3 P1 bytes per row, an STM-64 has 3×8 P1 bytes per row and an STM-256 has 3×32 P1 bytes per row.

- In the STM-64 and STM-256 overhead the byte [9, (3N + 2)] is allocated to extend the M1 capability; this is referred to as M0.
- In the STM-256 overhead 144 bytes have been allocated to provide an extra data communications channel with a larger bandwidth: D13 ... D156, i.e., a 9216 kbit/s capacity.

3.1.2   AUG-N Structure

Existing *Administrative Unit Group* (AUG) structures are AUG-1, AUG-4, AUG-16, AUG-64, and AUG-256.

The AUG-N fits exactly into the payload area of an STM-N and can be depicted as shown in Figure 3-6 where $X = 9 \times N$ and $Y = 270 \times N$ for $N = 1$, 4, 16, 64, and 256. The area [4, 1 ... X] is used to transport the AU-n pointers and the area [1 ... 9, (X + 1) ... Y] is used to transport the AU-n payload.

The AUG-N structure can be used to map a single *contiguous concatenated Administrative Unit of level N* (AU-4-Nc) or to multiplex several lower level AUG-Ns.

An AUG-256 can contain a single AU-4-256c or four AUG-64, an AUG-64 can contain a single AU-4-64c or four AUG-16, an AUG-16 can contain a single AU-4-16c or four AUG-4, an AUG-4 can contain a single AU-4-4c or four AUG-1, and an AUG-1 can contain a single AU-4 or three AU-3.

Due to the contiguous concatenation the multiplexing of an AUG-N into a higher-level AUG-M (M > N) is N column-interleaved. For example, when multiplexing four AUG-4 into an AUG-16 four columns of the first AUG-4 will be mapped into the AUG-16 structure followed by four columns of the second AUG-4, followed by four columns of the third AUG-4, followed by four columns of the fourth AUG-4, followed by four columns of the first AUG-4, etc.

3.1.3   AU-n Structure

Existing *Administrative Unit* (AU) structures are AU-3, AU-4, AU-4-4c, AU-4-16c, AU-4-64c, and AU-4-256c.

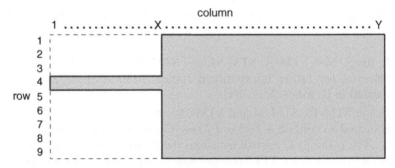

Figure 3-6. AUG structure

An AU-n structure consists of three elements:

- The payload area, which is mapped directly into the corresponding AUG-N or which is column-interleave multiplexed with other AU-n into a higher-level AUG-N.
- The VC-n that will be transported by the AU-n. This VC-n "floats" in the payload area of the AU-n to accommodate frequency differences between the VC-n clock and the AU-n clock.
- A pointer that indicates the offset of the first byte of the VC-n relative to the AU-n payload area. The pointer process is described in Chapter 6.

Figure 3-7 shows how an AU-4 (light gray) is mapped into an AUG-1 (dark gray), and the "floating" VC-4.

Figure 3-8 shows the AU-3 structure and the VC-3 {STS-1 SPE} it transports. Because the AU-3 is designed to map directly into an STM-0 {OC-1} it has a payload area of 87 columns.

To map a VC-3 {STS-1 SPE}, which has 85 columns, in this area, two columns containing fixed stuff (dark gray) have to be added to the VC-3 frame structure. Three AU-3 can also be column-interleave multiplexed into an AUG-1. This is shown schematically in Figure 3-9, only the byte-interleave multiplexing of the AU-3 pointer is depicted.

3.1.4   TUG-n Structures

Existing *Tributary Unit Group* (TUG) structures are TUG-3 and TUG-2. The TUG-3 is a substructure of a VC-4. Three TUG-3 can be column-interleave multiplexed into a VC-4.

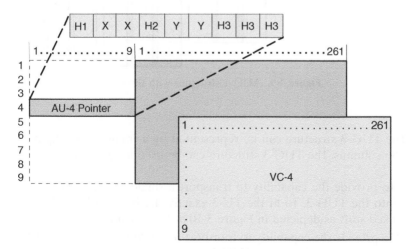

Figure 3-7. AUG-1 structure with AU-4 and VC-4

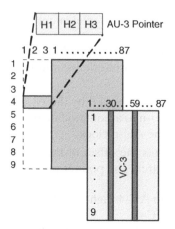

Figure 3-8. AU-3 structure with VC-3

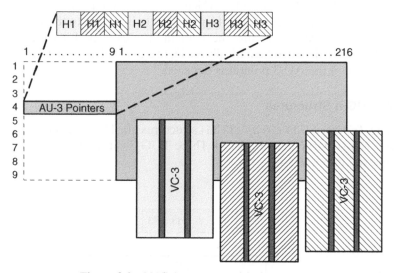

Figure 3-9. AUG-1 structure with three AU-3

The TUG-3 structure can be represented by a frame consisting of 9 rows and 86 columns. The TUG-3 structure can be used in two ways:

- To provide the capability to transport a single VC-3, a TU-3 is mapped into the TUG-3. To fit the TU-3 exactly the bytes in [4 … 9, 1] contain fixed stuff as depicted in Figure 3-10 at the left side.
- To provide the capability to transport multiple VC-2, VC-12 or VC-11 by substructuring, i.e., seven TUG-2 structures are column-interleave multi-

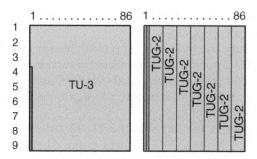

Figure 3-10. TUG-3 strucute

plexed into the TUG-3. To fit the seven TUG-2 exactly, all bytes of the first two columns of the TUG-3 contain fixed stuff as depicted in Figure 3-10 at the right side. (Note: *Figure 3-10 does NOT show the column-interleave multiplexing.*)

The TUG-2 structure can be represented by a frame consisting of 9 rows and 12 columns. The TUG-2 structure can be used in three ways:

- To provide the capability to transport a single VC-2, a TU-2 is mapped into the TUG-2. The TU-2 fits exactly in the TUG-2 structure.
- To provide the capability to transport multiple VC-12, three TU-12s are column-interleave multiplexed into the TUG-2.
- To provide the capability to transport multiple VC-11, four TU-11s are column-interleave multiplexed into the TUG-2.

3.1.5 TU-n Structures

Existing *Tributary Unit* (TU) structures are TU-11, TU-12, TU-2, and TU-3. The equivalent SONET structures are VT1.5, VT2, and VT6. The TU-3 does not exist in the SONET multiplex structure.

A TU-n structure consists of three elements:

- The payload area, which is mapped directly into the corresponding TUG-n or which is column-interleave multiplexed with other TU-n into a higher level TUG-n.
- The VC-n transported by the TU-n. This VC-n "floats" in the payload area of the TU-n to accommodate frequency differences between the VC-n clock and the TU-n clock.
- A pointer that indicates the offset of the first byte of the VC-n relative to the TU-n payload area. The pointer process is described in Chapter 6.

Figure 3-11 shows how a TU-3 (light gray) is mapped into a TUG-3 (dark gray), and the "floating" VC-3.

In the TUG-2 only one byte at [1, 1] is available for the TU-2 pointer, the four bytes of the TU-2 pointer are transmitted in successive frames; this means that a complete TU-2 pointer is repeated every 500 μsec.

Figure 3-12 shows the TU-2 structure (dark gray) with its pointer and the "floating" VC-2 (light gray).

Figure 3-13 shows at the left side the TU-12 structure (dark gray) with its pointer and the "floating" VC-12 (light gray). At the right side it shows the TU-11 structure (dark gray) with its pointer and the "floating" VC-11 (light gray). In the TUG-2 structure only one byte is available for the TU-12 and TU-11 pointer, the four pointer bytes are transmitted in successive frames; this means that a complete TU-12 and TU-11 pointer is repeated every 500 μsec.

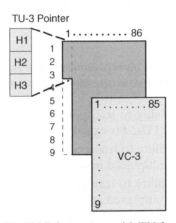

Figure 3-11. TUG-3 structure with TU-3 and VC-3

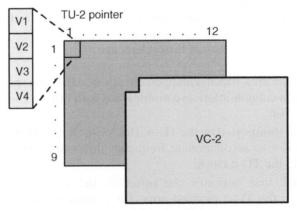

Figure 3-12. TUG-2 structure with TU-2 and VC-2

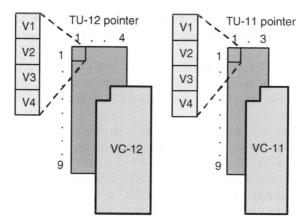

Figure 3-13. TU-12 structure and TU-11 structure

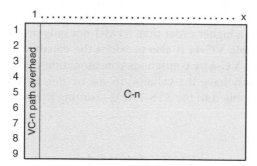

Figure 3-14. VC-n frame structure

Three TU-12 structures are column-interleave multiplexed into a TUG-2 structure.

Four TU-11 structures are column-interleave multiplexed into a TUG-2 structure.

3.1.6 VC-n Structures

Existing *Virtual Container* (VC) frame structures are VC-11, VC-12, VC-2, VC-3, and VC-4. The equivalent SONET frame structures are *Synchronous Payload Envelopes* (SPE): VT1.5 SPE, VT2 SPE, VT6 SPE, STS-1 SPE, and STS-3c SPE.

In general these frame structures are referred to as VC-n for the higher order VCs and VC-m for the lower order VCs. SDH considers VC-4 higher order, SONET considers VC-3 {STS-1 SPE} and VC-4 {STS-3c SPE} higher order.

For the VC-3 {STS-1 SPE} and VC-4 {STS-3c SPE} the frame structure can be represented by a frame consisting of 9 rows and x columns, as shown in Figure 3-14, for VC-3 x = 85 and for VC-4 x = 261. Column 1 contains the VC-n path overhead as described in Section 4.2.6 of Chapter 4, the Sn specific path overhead. Columns 2 ... x provide the capability to transport payload container C-n.

For the VC-11 {VT1.5 SPE}, VC-12 {VT2 SPE}, and VC-2 {VT6 SPE} the frame structure can be represented by a frame consisting of 9 rows and y columns except byte [1, 1], as shown in Figure 3-15. For VC-11 y = 3, for VC-12 y = 4, and for VC-2 y = 12.

An alternative way to depict a VC-m is shown in Figure 3-16 where the frame consists of 4 rows and z columns and the frame rate is 500 μsec. For VC-11 z = 26, for VC-12 z = 35, and for VC-2 z = 107.

In the VC-m only one byte [1, 2] is available for the *VC-m Path OverHead* (VC-m POH), and the four bytes of the VC-m POH are transmitted in successive frames; this means that a complete VC-m POH is repeated every 500 μsec. The VC-m POH is described in Section 4.3.6 of Chapter 4, the Sm specific path overhead. The remaining area provides the capability to transport payload container C-m. The byte in [1, 1] is omitted to be able to fit the VC-m exactly into its related TU-m (see Figures 3-12 and 3-13).

Multiplexing to a higher order than STM-1 not only provides the capability to transport multiple VC-4s, it also provides the capability to transport containers larger than VC-4 by contiguous concatenation creating a VC-4-Xc. In this case X can only have the value 4, 16, 64, or 256. In SONET contiguous concatenation is applied to the STS-1 SPE resulting in the STS-Nc SPE struc-

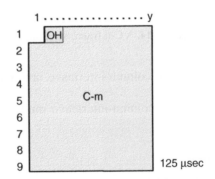

Figure 3-15. VC-m frame structure

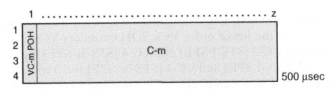

Figure 3-16. VC-m frame structure

ture with N = 3, 12, 48, 192, and 768. Because the contiguous concatenation is applied to column-interleave multiplexed VC-4s column 1 contains the VC-4-Xc path overhead and the following (X-1) columns, that contained the path overhead of the other VC-4, are filled with fixed stuff.

The VC-4-Xc frame structure can be represented by a frame consisting of 9 rows and 261 × X columns as depicted in Figure 3-17, the (X-1) columns fixed stuff are dark gray.

To provide more payload capacity to map ATM cells the VC-2-Xc is defined (X = 1 ... 7).

The VC-2-Xc frame structure can be represented by a frame consisting of 9 rows and 12 × X columns except byte [1, 1] as depicted in Figure 3-18. It contains the VC-2-Xc path overhead byte, the (X-1) fixed stuff bytes (dark gray) and the payload container bytes C-2-X.

An alternative way to depict a VC-2-Xc is shown in Figure 3-19 where the frame consists of 4 rows and 107 × X columns and the frame rate is 500 μsec.

To provide more flexibility and more granularity in the available bandwidth *virtual concatenated VCs* (VC-n-Xv) are defined. Existing VC-n-Xv are VC-11-Xv, VC-12-Xv, VC-2-Xv, VC-3-Xv, and VC-4-Xv.

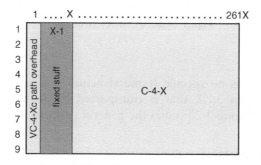

Figure 3-17. VC-4-Xc frame structure

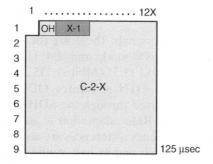

Figure 3-18. VC-2-Xc frame structure

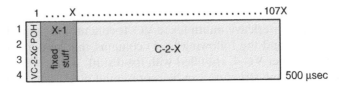

Figure 3-19. VC-2-Xc frame structure

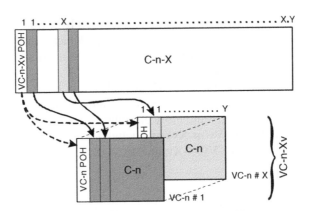

Figure 3-20. VC-n-Xv structure

The VC-n-Xv has no specific frame structure, it is actually a C-n-X, as described in Section 3.1.7, that is transported by X individual VC-ns, as described above. Figure 3-20 shows the general structure.

3.1.7 Payload Containers C-n

In SDH the Container C-n is the frame structure that is used to accommodate all client or tributary signals. An overview of all existing C-n is provided in Table 3-2.

Several types of client signals can be mapped into the C-n payload area:

- *Constant Bit-Rate* (CBR) signals. These are the PDH framed signals E1 (2.048 Mbit/s), E3 (34.368 Mbit/s), and E4 (139.264 Mbit/s), and the SONET framed signals DS1 (1.554 Mbit/s), DS2 (6.312 Mbit/s), and DS3 (44.736 Mbit/s). Also the OTN structures ODU1 and ODU2 can be mapped into and transported through the SDH network by a VC-4-17v, respectively, a VC-4-68v. Rate adaptation is accomplished by inserting fixed stuff columns. Frequency differences are accommodated by bit stuffing for asynchronous signals and by byte stuffing for synchronous signals as described in Chapter 6.

TABLE 3-2. C-n sizes and bit-rates

C-n	Rows	Columns	Bytes	X	Frame-rate	Bit-rate in kbit/s
C-11	4	25	100		500 µsec	1,600
C-11-X	4	X × 25	X × 100	1 ... 64	500 µsec	1,600 ... 102,400
C-12	4	34	136		500 µsec	2,176
C-12-X	4	X × 34	X × 100	1 ... 64	500 µsec	2,176 ... 139,264
C-2	4	106	424		500 µsec	6,784
C-2-X	4	X × 106	X × 100	1 ... 64	500 µsec	6,784 ... 434,176
C-3	9	84	756		125 µsec	48,384
C-3-X	9	X × 84	X × 100	1 ... 256	125 µsec	48,384 ... 12,386,304
C-4	9	260	2,340		125 µsec	149,760
C-4-X	9	X × 260	X × 100	1 ... 256	125 µsec	149,760 ... 38,338,560
C-4-4c	9	1,040	9,360		125 µsec	599,040
C-4-16c	9	4,160	37,440		125 µsec	2,396,160
C-4-64c	9	16,640	149,760		125 µsec	9,584,640
C-4-256c	9	66,560	599,040		125 µsec	38,338,560

- *Asynchronous Transfer Mode* (ATM) cells are mapped consecutively and use the full C-n bandwidth. Rate adaptation is accomplished as part of the cell stream creation process by either inserting idle cells or by discarding cells as described in [ITU-T Rec. I.432.1].
- Packet-based signals are mapped using the *Generic Framing Procedure* (GFP) and can fill the whole C-n frame. Rate adaptation is accomplished by inserting GFP "Idle" frames. Refer to [ITU-T Rec. G.7041].

More detailed mapping information is provided in the description of the VC-n adaptation functions in Chapter 4, Section 4.2, on the Sn path layer and in Section 4.3, on the Sm path layer.

3.2   OTN FRAMES AND STRUCTURES

In OTN the following frames and structures can be distinguished: OTUk, ODUk, and OPUk. The standard that specifies the OTN structures is [ITU-T Rec. G.709].

Currently three levels of OTN multiplex are defined (k = 1, 2, and 3). Figure 3-21 shows the relation between these frame structures.

All OTN frames are based on a generic OTN frame structure consisting of 4 rows and 3,824 columns, as shown in Figure 3-22. Multiplexing is achieved by increasing the frame repetition rate X.

The separate frame structures are described in the following sections.

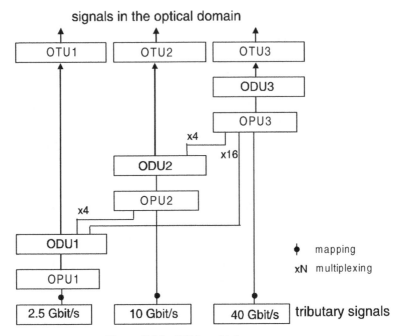

Figure 3-21. OTN multiplex structure

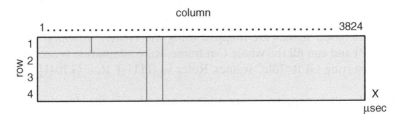

Figure 3-22. Generic OTN frame structure

3.2.1 OTUk Frame Structure

The *Optical channel Transport Unit of level k* (OTUk) consists of the generic frame structure with an extension of 256 columns. The extension contains the *Forward Error Correction* (FEC) code. The FEC can be used to detect bit errors and then has the capability to detect up to 16 symbol errors. The FEC can also be used to correct bit errors and then has the capability to correct up to 8 symbol errors.

The bytes in [1, 8 ... 14] contain the OTUk-specific overhead (OTUk OH). The bytes in [1, 1 ... 7] contain the *Frame Alignment Overhead* (FA OH) required to indicate the start of frame in the serial bit stream.

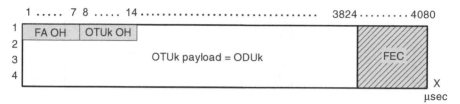

Figure 3-23. OTUk frame structure

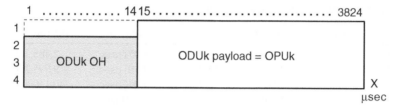

Figure 3-24. ODUk frame structure

The complete OTUk frame structure consists of 4 rows and 4,080 columns as shown in Figure 3-23. The OTUk payload area is used to transport one ODUk.

3.2.2   ODUk Frame Structure

The *Optical channel Data Unit of level k* (ODUk) consists of the generic frame structure except the bytes in [1, 1 ... 14]. The bytes in [2 ... 4, 1 ... 14] contain the ODUk-specific overhead (ODUk OH).

The complete ODUk frame structure is shown in Figure 3-24. The ODUk payload area is used to transport one OPUk.

3.2.3   OPUk Frame Structure

The *Optical channel Payload Unit of level k* (OPUk) consists of the generic frame structure except the bytes in [1 ... 4, 1 ... 14]. The bytes in [1 ... 4, 14, 15] contain the OPUk-specific overhead (OPUk OH).

The complete OPUk frame structure consists of 4 rows and 3,810 columns as shown in Figure 3-25. The first two columns (15 and 16) are dedicated to the OPUk overhead (OPUk OH). The OPUk is used to transport lower level ODUk frame structures or OTN client signals.

Table 3-3 contains an overview of the bandwidth that is provided by each of the OPUk frame structures. **Note**: *The bit-rates are nominal; the accuracy of the OTN signals is ±20 ppm.*

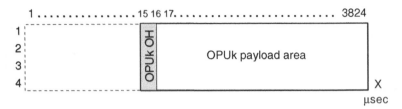

Figure 3-25. OPUk frame structure

TABLE 3-3. OPUk payload sizes and bit-rates

OPUk	Rows	Columns	Bytes	Frame-rate	Bit-rate in kbit/s
OPU1	4	3,808	15,232	48.971 µsec	2,488,320.000
OPU2	4	3,808	15,232	12.191 µsec	9,995,276.962
OPU3	4	3,808	15,232	3.035 µsec	40,150,519.322
OPU4	4	3,808	15,232	~1.0 µsec	~100 Gbit/s

Currently (12-2007) ITU-T is defining the OTU4/ODU4/OPU4 frame structures in cooperation with the IEEE. The objective is to define a frame structure that will efficiently transport the new 100 Gbit/s Ethernet signal.

Several types of client signals can be mapped into the OPUk payload area:

- Lower level ODUk frame structures can be multiplexed. An OPU2 can transport 4 ODU1 and an OPU3 can transport up to 4 ODU2 and/or 16 ODU1. The transport capability of the new OPU4 is currently studied.
- *Constant Bit-Rate* (CBR) signals, e.g., the SDH structures STM-16, STM-64, and STM-256 can be mapped as CBR2G5, CBR10, and CBR40 signals into, respectively, the OPU1, OPU2, and OPU3. Rate adaptation is accomplished by inserting fixed stuff columns and byte justification. The justification process is described in Section 5.3, positive and negative justification.
- *Asynchronous Transfer Mode* (ATM) cells can utilize the full OPUk bandwidth. Rate-adaptation is accomplished as part of the cell stream creation process by either inserting idle cells or by discarding cells. Refer to [ITU-T Rec. I.432.1].
- Packet-based signals are mapped using the *Generic Framing Procedure* (GFP). Rate adaptation is accomplished by inserting GFP Idle frames. Refer to [ITU-T Rec. G.7041].

More detailed information is provided in the description of the OTN OPUk adaptation functions in Chapter 4, Section 4.10.

Also, in the OTN more flexibility and more granularity in the available bandwidth can be provided by *virtual concatenated OPUks* (OPUk-Xv).

The OPUk-Xv provides the capability to transport the payload of X OPUks. This payoad is transported by X individual OPUks similar to the transport of a VC-n-Xv. Figure 3-26 shows the general structure. The VCOH contains the VCAT overhead information and the PTI indicates the virtual concatenation.

3.3 PDH FRAME STRUCTURES

This section describes the most common PDH frame structures. It is limited to those frame structures that are currently defined as client signals of SDH and SONET.

3.3.1 E11 (DS1)—1.544 Mbit/s

The E11 frame structure is defined in [ITU-T Rec. G.704] and consists of 193 bits; the frame is repeated every 125 μsec. The first bit of the frame, i.e., the F-bit, is used for frame alignment, performance monitoring, and provides a data channel of 4 kbit/s. The following 192 bits provide 24 octets; each octet has the capability to transport a 64 kbit/s channel. Figure 3-27 shows the E11 frame structure.

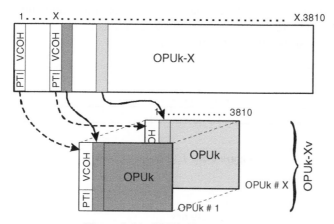

Figure 3-26. OPUk-Xv structure

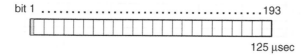

Figure 3-27. E11 frame structure

E11 frames are mapped bit-asynchronous into the SDH container C-11 and bit-justification is used to compensate frequency differences.

3.3.2 E12 (E1)—2.048 Mbit/s

The E12 frame structure is defined in [ITU-T Rec. G.704] and consists of 256 bits; the frame is repeated every 125 μsec. The 256 bits provide 32 octets or timeslots. The first timeslot is used for frame alignment, performance monitoring and provides a data channel of 4 kbit/s. The remaining 31 timeslots have the capability to transport a 64 kbit/s channel. Figure 3-28 shows the E12 frame structure.

E12 frames can be mapped bit-asynchronous into the SDH container C-12 and bit-justification is used to compensate frequency differences. E12 frames can also be mapped byte-synchronous into the SDH container C-12; in this case pointer processing is used to compensate frequency differences (see Chapter 6).

3.3.3 E21 (DS2)—6.312 Mbit/s

The E21 frame structure is defined in [ITU-T Rec. G.743] and consists of 1,176 bits, the frame is repeated every 186.3 μsec. The frame is divided into four subframes and each subframe is further divided into 6 blocks of 49 bits. The first bit of each block will carry overhead information. The remaining 48 bits of each block are used to bit-interleave multiplex four E11 signals. Together the 24 overhead bits in a DS2 frame are used for frame and subframe alignment of the DS2 and for positive bit-justification control.

Figure 3-29 shows the E21 frame structure.

Figure 3-28. E12 frame structure

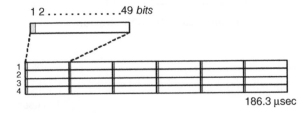

Figure 3-29. E21 frame structure

E21 frames are mapped bit-asynchronous into the SDH container C-2 and bit-justification is used to compensate frequency differences.

3.3.4   E31 (E3)—34.368 Mbit/s

The E31 frame structure as defined in [ITU-T Rec. G.753] consists of 2,148 bits; the frame is repeated every 62.5 μsec. In the frame 12 bits are allocated for frame alignment, service functions, and national use (grayed); 16 bits are used for bit-justification control and opportunity (hatched) of the four multi-plexed tributary E22 (E2) signals. The tributary signals are bit-interleave mul-tiplexed in the remaining 2,120 bits. **Note:** *[ITU-T Rec. G.753] defines the E31 multiplex with positive and negative justification, there is also [ITU-T Rec. G.751] that defines the E31 with positive justification only.*

Figure 3-30 shows the E31 frame structure; for completeness the frame structure of the E22 (E2) 8.448 Mbit/s is also shown.

E31 frames are mapped bit-asynchronous into the SDH container C-3 and bit-justification is used to compensate frequency differences.

3.3.5   E32 (DS3)—44.736 Mbit/s

The E32 frame structure is defined in [ITU-T Rec. G.752] and consists of 4,760 bits; the frame is repeated every 106.4 μsec. The frame is divided into seven subframes and each subframe is further divided into 8 blocks of 85 bits. The first bit of each block is allocated to carry overhead information. The remain-ing 84 bits of each block are used to bit-interleave multiplex seven E21 signals. Together the 56 overhead bits in a DS3 frame are used for frame and subframe alignment of the DS3 and for positive bit-justification control. Figure 3-31 shows the E31 frame structure.

E32 frames are mapped bit-asynchronous into the SDH container C-3 and bit-justification is used to compensate frequency differences.

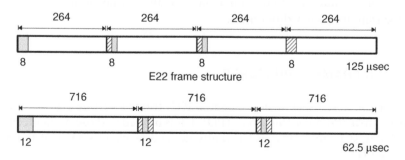

Figure 3-30. E31 frame structure

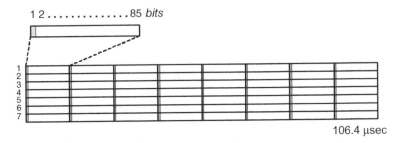

Figure 3-31. E32 frame structure

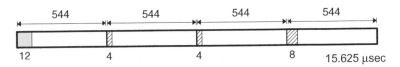

Figure 3-32. E4 frame structure

3.3.6 E4—139.264 Mbit/s

The E4 frame structure as defined in [ITU-T Rec. G.754] consists of 2,176 bits; the frame is repeated every 15.625 µsec. In the frame 12 bits are allocated for frame alignment and service functions (grayed), 16 bits are used for bit-justification control and opportunity (hatched) of the four multiplexed tributary E31 (E3) signals. The tributary signals are bit-interleave multiplexed in the remaining 2,148 bits.

Note: *[ITU-T Rec. G.754] defines the E4 multiplex with positive and negative justification. There is also [ITU-T Rec. G.751] that defines the E4 with positive justification only.*

Figure 3-32 shows the E4 frame structure. E4 frames are mapped bit-asynchronous into the SDH container C-4 and bit-justification is used to compensate frequency differences.

3.4 PTN FRAME STRUCTURES

This section describes the most common PTN frame structures; they are often referred to as *Protocol Data Units* (PDU). It is limited to those packet technology frame structures that are currently defined as client signals of the STN and the OTN.

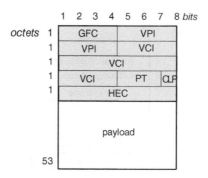

Figure 3-33. ATM frame structure

3.4.1 ATM Cell

The frame structure of an ATM cell is defined in [ITU-T Rec. I.361]. The ATM cell overhead consists of the following fields, depicted in Figure 3-33:

- The *Generic Flow Control* (GFC) field is used at the *User–Network Interface* (UNI); at the *Network–Node Interface* (NNI) it is part of the VPI.
- The Virtual Path Identifier (VPI) field and the Virtual Channel Identifier (VCI) field together provide a 24 bits UNI or 28 bits NNI routing field used for routing cells in a network.
- The Payload Type (PT) field is used to indicate that the cell is used either for user data or for OAM data.
- The Cell Loss Priority (CLP) field is used to indicate that the cell is drop eligible (if set to "1").
- The Header Error Control (HEC) field contains an error detecting and correction code to validate the cell header. Single bit errors can be corrected and multiple bit errors can be detected.

The ATM payload field is fixed at 48 octets.

3.4.2 Ethernet MAC PDU

The frame structure of an Ethernet MAC PDU is defined in [IEEE 802.3] Section 3.1. The Ethernet MAC frame overhead consists of the following fields, depicted in Figure 3-34:

- The preamble field of 62 bits that contain an alternating "0"/"1"pattern which is used to synchronize the receive clock

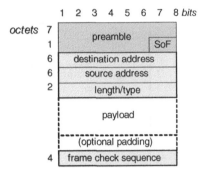

Figure 3-34. Ethernet MAC frame structure

- The *Start of Frame* (SoF) delimiter field of two bits that have the value "11."
- The *Destination Address* (DA) field consists of six octets that may be a unicast, a multicast or a broadcast address.
- The *Source Address* (SA) field consists of six octets and is set to the sender's globally unique node address.
- The length/type field consisting of two octets these are used to either indicate the length of the payload in octets, or to identify the type of protocol that is carried, i.e., the *Service Access Point* (SAP) information. The type field may also be used to indicate when a Tag field is added to the PDU.
- *Frame Check Sequence* (FCS) field at the end of the Ethernet MAC PDU. It is a four-octet field that contains error detection code. The code is calculated over all octets in the PDU except the preamble and SoF bits.

The Ethernet MAC payload field can by default contain up to 1,500 information octets, but other values are negotiable. Padding octets can be added to reach the maximum size.

3.4.3 MPLS-TP and MPLS PDU

The frame structure of an MPLS-TP and MPLS PDU, either unicast or multicast, is defined in [IETF RFC3032]. The MPLS-TP/MPLS frame overhead consists at least of one MPLS specific label stack entry of four octets with the following fields, depicted in Figure 3-35:

- The label field carries the actual value of the MPLS/ MPLS-TP label. The label value is used to retrieve either the next hop to which the PDU is to be forwarded, or the operation that has to be performed on the label stack before forwarding.

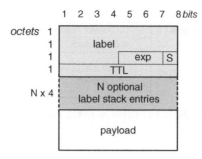

Figure 3-35. MPLS-TP/MPLS frame structure

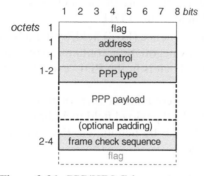

Figure 3-36. PPP/HDLC frame structure

- The exp field is reserved.
- The S bit indicates the bottom of the stack (if set to "1").
- The *Time To Live* (TTL) field is used to encode the lifetime of the PDU in the network.

The length of the MPLS-TP/MPLS payload information field depends on the length of the network layer technology PDUs.

3.4.4 HDLC/PPP PDU

The frame structure of a *Point to Point Protocol* (PPP) PDU is defined in [IETF RFC1661] Section 2. A *High-Level Data Link Control* (HDLC) like PDU is used to encapsulate HDLC/PPP payloads. The HDLC-like PDU is defined in [IETF RFC1662] Section 3, which specifies the following fields, depicted in Figure 3-36:

- The flag field is used as PDU delimiter. It has the value 0x7E, the flag at the end of a PDU may be used as the flag at the begin of the next PDU.

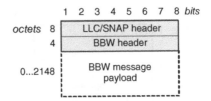

Figure 3-37. FC-BBW SONET frame structure

- The address field is set to 0xFF; the "All-Stations" address.
- The control field is set to 0x03; the "Unnumbered Information" command.
- The PPP type field, either one or two octets, identifies the datagram in the PPP payload field of the PDU.
- The Frame Check Sequence (FCS) field, by default two octets but extendable to four octets by negotiation, contains error detection code. The code is calculated over all octets in the PDU except the flag octets.

The PPP payload field contains of zero or more information octets. The maximum length of the PPP payload field including the padding is by default 1,500, but other lengths can be negotiated. Padding octets can be added to reach the maximum size.

3.4.5 FC-BBW_SONET PDU

The frame structure of a *Fibre Channel BackBone WAN* (FC-BBW) PDU is defined in [ANSI INCITS 342]. The following fields are specified, as shown in Figure 3-37:

- A fixed eight octet *Logical Link Control* (LCC)/*Sub Network Access Protocol* (SNAP) field, which indicates the payload type. The LLC consists of three octets, the *Destination Service Access Point* (DSAP), *Source Service Access Point* (SSAP), and *Control* (CTRL). The SNAP consists of five octets, a three octet Organizationally Unique Identifier (OUI) and a two octet Protocol Identification (PID).
- A four octet BBW header indicates the type of flow control.

The BBW payload area contains 0 ... 2,148 information octets.

CHAPTER 4

NETWORK MODELING: LAYERS AND ATOMIC FUNCTIONS

In this chapter the individual network layers are described that can be identified in a network. The basis for these descriptions is the functional model of the network as is described in Section 2.2 of Chapter 2.

Adjacent layers in a functional model have a client ↔ server relation, where the server layer transports one or more (multiplexed) client layer signals. In general the client layer is drawn on top of the server layer; this convention is used in this chapter, too.

Each network layer contains basically an adaptation function at the client layer side, a path or trail termination function, and a connection function at the server layer side.

This chapter describes the network models of SDH (and SONET) and OTN.

4.1 THE SYNCHRONOUS DIGITAL HIERARCHY SDH

Figure 4-1 shows the functional model of the SDH network; it is applicable to the SONET network as well, and it has the same structure. If necessary, differences between SDH and SONET are identified. This particular model is used to show the different layers that are identified in the SDH network; it does not represent an equipment model or a network model.

The ComSoc Guide to Next Generation Optical Transport: SDH/SONET/OTN,
by Huub van Helvoort
Copyright © 2009 Institute of Electrical and Electronics Engineers

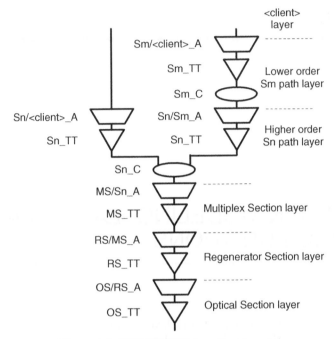

Figure 4-1. SDH/SONET functional model

The descriptions of each network layer that follow are ordered according to Figure 4-1 from the top to the bottom, except the Sn and Sm path layers. Within each network layer descriptions of the three atomic functions are provided. This description combines the frame specifications described in [ITU-T G.707] and the equipment specifications described in [ITU-T G.783].

4.2 THE SDH HIGHER ORDER PATH (Sn) LAYER

The higher order path layer Sn, where "n" denotes the *Virtual Container* (VC) type, is the layer that provides the connectivity in an SDH network.

The Sn layer can serve as the transport layer for either the lower order path layer, designated Sm to distinguish the two, or any non-SDH tributary signal layer that is mapped into the higher order VC-n. The higher order path layer Sn has the MSn section layer, described in Section 4.4, as its server layer. Figure 4-2 shows the generic Sn network layer functional model and the related frame structures.

This model is applicable to the SDH higher order S4 path layer as well as the SONET higher order S3 path layer as depicted in Figure 3-1 of Chapter 3.

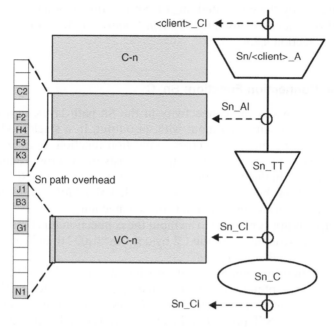

Figure 4-2. Sn network layer model

The Sn specific *Path OverHead* (Sn-POH) is used to transfer OAM information.

4.2.1   The Source Adaptation Function: Sn/<client>_A_So

The characteristic information of the <client> signal is mapped and/or multiplexed into the container C-n; the transport entity of the Sn path layer. Because there are several <client> signals possible, the specific adaptation functions Sn/<client>_A will be described separately after the generic description of the Sn atomic functions.

This function also inserts the Sn adaptation specific OAM in the Sn-POH, i.e., bytes C2, H4, K3, F2, and F3 described in Table 4-1 and shown in Figure 4-2.

4.2.2   The Source Termination Function: Sn_TT_So

This function inserts the Sn path specific OAM in the Sn-POH, i.e., the J1, B3, G1, and N1 bytes described in Table 4-2 and shown in Figure 4-2. The API received from the *Element Management Function* (EMF) is inserted in the J1 byte. The BIP-8 code is calculated and inserted in the B3 byte. The bits in the

G1 byte are set by the co-located Sn_TT_Sk function. Normally the N1 byte is set to all-zero; it can be used for *Tandem Connection Monitoring* (TCM) as described in Section 4.2.9.

4.2.3   The Connection Function: Sn_C

This function provides the connectivity in the Sn path layer. It is a matrix function that can connect an input with an output. In a single node the Sn connection function models the cross-connection function. In a network the connectivity, i.e., the network connection, consists of the cross-connect functions in the nodes; in the model the *Sub-Network Connection* (SNC) functions and the connections between the nodes modeled as a *Link Connection* (LC). Figure 4-3 shows an example of a network connection.

If an output is not connected to an input the connection function will generate an Sn frame structure with the C2 byte set to "0x00" to indicate the open connection, i.e., "*path unequipped.*"

The connection function Sn_C can provide the protection of the Sn trails as described in Chapter 7. Protection switching can be initiated by the *Server Signal Fail* (SSF) and *Server Signal Degrade* (SSD) conditions received from the server layer. An SSF present at the input is transferred to the same output as the payload data.

4.2.4   The Sink Termination Function: Sn_TT_Sk

Extracts the Sn path specific OAM from the Sn-POH, i.e., the J1, B3, G1, and N1 bytes described in Table 4-2 and shown in Figure 4-2. This information is used for checking the Sn layer connectivity, the Sn path status, and its performance.

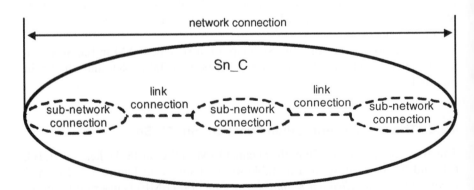

Figure 4-3. Network, sub-network and link connections

If the API in the J1 byte is not equal to the *Expected Trail Identifier* (ExTI), this will be reported to the EMF as *Trail Information Mismatch* (TIM) defect. If the C2 byte contains the value "0x00" the *Unequipped* (UNEQ) defect is detected. If either a TIM defect or an UNEQ defect is detected or an SSF is received from the server layer, a *Trail Signal Fail* (TSF) is sent to the succeeding adaptation function and a *Remote Defect Indication* (RDI) is sent to the co-located Sn_TT_So function for insertion in the G1 byte. A received RDI is sent to the EMF to be used for fault handling. If a TIM or UNEQ defect is detected the output signal is replaced by an *Alarm Indication Signal* (AIS), i.e., an all-ones signal. The received value of B3 is compared to the calculated BIP-8. The number of detected bit-errors is used for near-end performance monitoring and sent as *Remote Error Indication* (REI) to the co-located Sn_TT_So function for insertion in the G1 byte. It is also used to detect the *Signal Degrade* (SD) defect. This defect is transferred to the succeeding adaptation function as *Trail Signal Degrade* (TSD). The received REI is used for local far-end performance monitoring. The value of the N1 byte is ignored unless it is used for TCM as described in Section 4.2.9.

4.2.5 The Sink Adaptation Function: Sn/<client>_A_Sk

Recovers or de-multiplexes the characteristic information of the <client> signal from the container C-n. The function will also extract the Sn adaptation specific OAM from the Sn-POH, i.e., the bytes C2, H4, K3, F2, and F3 described in Table 4-1 and shown in Figure 4-2. The pointer process will report a *Loss of Pointer* (LOP) defect. If the value of the C2 byte does not match the mapping technology supported by this adaptation function a *PayLoad Mismatch* (PLM) defect is detected. If the H4 byte is used for multiframe alignment it can detect a *Loss of Multiframe* (LOM) defect. If a LOP or PLM or LOM defect is detected the output signal is replaced by an AIS (all-ones) signal and an SSF signal is sent to the client layer.

4.2.6 The Sn Specific Path Overhead: Sn-POH

The next two tables describe the Sn-POH bytes.

Table 4-1 contains the overhead bytes dedicated to the mapping and/or multiplexing functionality of the Sn adaptation functions. Table 4-2 contains the overhead bytes dedicated to the path supervision functionality of the Sn trail termination functions.

4.2.7 The Client Specific Adaptation Functions in the Sn Path Layer

The S4 path layer can have the S3 path layer as its <client> signal layer. Both S4 and S3 path layers can have path layer S2, S12, or S11 as their <client> layer, as well as a non-SDH structured signal layer. Each of these possibilities is described in the following sections.

TABLE 4-1. VC-n (n = 3, 3-Xc, 4, 4-Xc) adaptation overhead bytes

Byte	Description
C2	*Path Signal Label* (PSL) byte—specifies the mapping technology used for mapping client signals in the VC-n. Examples of (hexadecimal) values: 0x00 unequipped—open connection in the connection function 0x02 TUG structured—the VC-n payload area is sub-structured 0x12 asynchronous client mapping of E4 into a VC-4 0x20 mapping of OTN ODUk 0x1A 10 Gbit/s Ethernet mapping 0x1B GFP [ITU-T G.7042] mapping—as described in Chapter 8 0xFF VC-AIS—indication of path AIS A complete list can be found in Table 9 of [ITU-T G.707],
F2	*Path User channel* byte—used for user communication between path elements. This channel can be used as a 64 kbit/s clear channel and have its own interface on the equipment.
H4	*Multiframe and Sequence Indicator* byte—used as a multiframe indicator for VC-2/VC-12/VC-11 payloads or provides a 16 byte control packet for VC-3/VC-4 virtual concatenation (see also Chapter 9).
F3	*Path User channel* byte—has the same purpose as the F2 byte. In SONET this byte is Z3—*growth* (for future use)
K3[1 ... 4]	Four *Automatic Protection Switching* (APS) signaling bits—allocated for protection switching at VC-4/VC-3 path levels.
K3[5 ... 8]	Four undefined bits—reserved for future use, ignored at reception. In SONET the K3 byte is Z4—*growth* (for future use)

4.2.7.1 The S3 Client to S4 Server Adaptation Function S4/S3_A

Figure 4-4 shows S4/S3_A atomic function and the related frame structures.

The VC-3 signal (85 columns) is mapped into a *Tributary Unit of level 3* (TU-3) and uses the TU-3 pointer (bytes H1, H2 and H3 described in Table 4-8) to enable the VC-3 to "float" in the TU-3 structure. The pointer process is described in Chapter 6. Fixed stuff bytes are added to the TU-3 to fit it into a *TU Group of level 3* (TUG-3) (86 columns). Three TUG-3 structures are byte interleave multiplexed into the payload area of a VC-4.

The following sections describe in detail the operation of the separate functions present in the functional model.

4.2.7.1.1 The Source Adaptation Function: S4/S3_A_So   This function multiplexes the VC-3 signals into the VC-4 structure. The C2 byte in the S4-POH is set to "0x02": the VC-4 is "TUG-structured."

TABLE 4-2. VC-n (n = 3, 3-Xc, 4, 4-Xc) termination overhead bytes

Byte	Description
J1	Higher-Order path *Access Point Identifier* (API) byte—used to transfer a 16-byte string as defined in clause 3 of [ITU-T Rec. G.831]. It is transmitted repetitively. A 64-byte string is also permitted for this API.
B3	*Path Error Detecting Code* (EDC) byte—this byte contains an 8 bit Path Bit Interleaved Parity code (BIP-8), even parity, to determine if a transmission error has occurred in a path. The BIP-8 is calculated over all bits of the previous virtual container and placed in the B3 byte of the current frame.
G1	*Path status* byte—This byte is used to convey the path terminating status and performance back to the path trail termination source function.

1	2	3	4	5	6	7	8
		REI		RDI		reserved	Spare

- Bits 1 ... 4: Path Remote Error Indication (REI)—set to the total number of errors that were detected by the path BIP-8 checker.
- Bit 5: Path Remote Defect Indication (RDI)—set to one if a server signal failure or trail signal failure is detected.
- Bits 6, 7: reserved bits—used only in SONET together with bit 5 to convey the Enhanced Remote Defect Indication (ERDI).
- Bit 8: spare bit—for future use

Byte	Description
N1	*Network operator* byte—allocated to support a Higher-Order Tandem Connection Monitoring (HO-TCM) function. It provides the capability for near-end and far-end performance monitoring of the TCM section. It also transfers the status of the incoming signal from the TCM source to the sink.

4.2.7.1.2 The Sink Adaptation Function: S4/S3_A_Sk This function demultiplexes the VC-3 signals from the VC-4 structure. The C2 byte extracted from the S4-POH shall have the value "0x02," if a different value is received the adaptation function cannot process the payload of the received VC-4 and reports to the EMF a PLM, consequently it will output an AIS signal and an SSF will be sent to the succeeding S3 connection function S3_C.

4.2.7.2 The Generic Sm Client to S4 Server Adaptation Function

S4/Sm_A This function multiplexes the lower order Sm (m = 2, 12, 11) path layer signals into the higher order S4 path layer for transport in the SDH network.

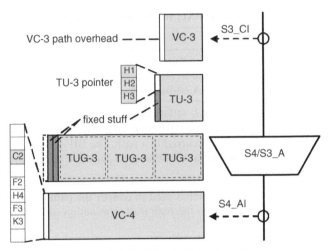

Figure 4-4. VC-3 to VC-4 adaptation

In the following sections the generic source and sink Sm to S4 adaptation functions are described followed by the S2, S12, and S11 specific adaptation functions.

4.2.7.2.1 The Source Adaptation Function: S4/Sm_A_So This function multiplexes the VC-m (m = 2, 12, 11) signals into the VC-4 structure. The C2 byte in the S4-POH is set to "0x02": the VC-4 is "*TUG-structured.*"

4.2.7.2.2 The Sink Adaptation Function: S4/Sm_A_Sk This function demultiplexes the VC-m (m = 2, 12, 11) signals from the VC-4 structure. If the C2 byte extracted from the S4-POH does not have the value "0x02" the adaptation function cannot process the payload of the received VC-4 and reports to the EMF a PLM defect, consequently it will output an AIS signal and an SSF will be sent to the Sm client path layer.

4.2.7.3 The S2 Client to S4 Server Adaptation Function S4/S2_A

Figure 4-5 shows the S4/S2_A atomic function and the related frame structures.

The VC-2 {VT6 SPE} signal is mapped into a TU-2 structure and uses the TU-2 pointer (bytes V1, V2, V3) to enable the VC-2 to "float" in the TU-2. The pointer process is described in Chapter 6. The TU-2 is mapped into a TUG-2 {VT group} structure and seven TUG-2s are byte interleave multiplexed into a TUG-3 structure. Finally three TUG-3s are byte interleave multiplexed into a VC-4 frame structure.

The adaptation function S4/S2_A is described in Section 4.2.7.2 with m = 2.

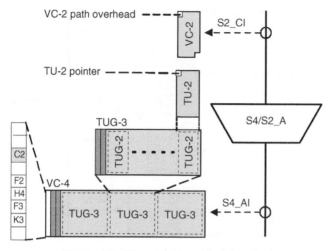

Figure 4-5. S2 to S4 adaptation function

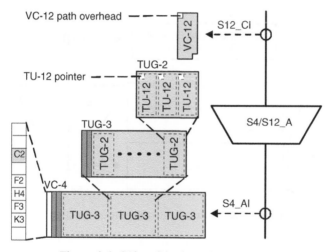

Figure 4-6. S12 to S4 adaptation function

4.2.7.4 *The S12 Client to S4 Server Adaptation Function S4/S12_A*

Figure 4-6 shows the S4/S12_A atomic function and the related frame structures.

The VC-12 signal is mapped into a TU-12 structure and uses the TU-12 pointer (bytes V1, V2, V3) to enable the VC-12 to "float" in the TU-12. Three TU-12 structures are byte interleave multiplexed into a TUG-2.

The adaptation function S4/S12_A is described in Section 4.2.7.2 above with m = 12.

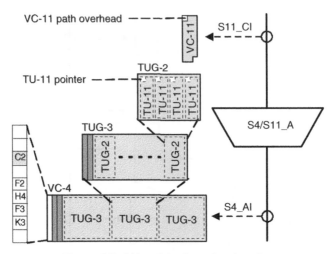

Figure 4-7. S11 to S4 adaptation function

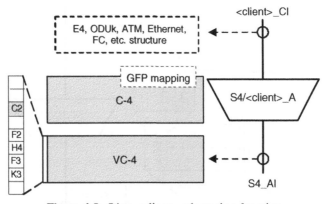

Figure 4-8. S4 to <client> adaptation function

4.2.7.5 The S11 Client to S4 Server Adaptation Function S4/S11_A

Figure 4-7 shows the S4/S11_A atomic function and the related frame structures.

The VC-11 signal is mapped into a TU-11 structure and uses the TU-11 pointer (bytes V1, V2, V3) to enable the VC-11 to "float" in the TU-11. Four TU-11 structures are byte interleave multiplexed into a TUG-2.

The adaptation function S4/S11_A is described in Section 4.2.7.2 with m = 11.

4.2.7.6 The Non-SDH Client to S4 Server Adaptation Function   Figure

4-8 shows the S4/<client>_A atomic function and the related frame structures.

Examples of Non-SDH signals are:

- A PDH E4 140Mbit/s signal. The C2 byte is set to "0x12."
- An OTN ODUk signal. The C2 byte is set to "0x20."
- ATM cells. The C2 byte is set to "0x13."
- packet frames, e.g., Ethernet or FC PDUs. The C2 byte is set to "0x1B".

The mapping methodologies are described in Chapter 8.

4.2.7.6.1   The Source Adaptation Function: S4/<client>_A_So   Maps the non-SDH signal structures into the C-4 frame structure as listed above. It also inserts the proper signal label into the C2 byte of the S4-POH. Because protection switching of non-SDH signals is not supported the K3 byte is not used.

4.2.7.6.2   The Sink Adaptation Function: S4/<client>_A_Sk   Recovers the non-SDH signal structures from the C-4 frame structure. If the C2 byte extracted from the S4-POH does not contain the proper signal label the adaptation function cannot de-map the payload of the VC-4 and reports to the EMF a PLM defect, consequently it will output either an all-ones signal or, in case of GFP mapping, a packet technology specific signal fail indication. The K3 byte information is ignored.

4.2.7.7   The Generic Sm Client to S3 Server Adaptation Function S3/Sm_A
This function adapts the lower order Sm ($m = 2, 12, 11$) path layer signals into the higher-order S3 path layer for transport in the SONET network. The functionality of the S3/Sm_A function is the same as the S4/Sm_A function apart from the mapping of the TUG-2. In the S3/Sm_A the TUG-2 is mapped into a VC-3 {STS-1-SPE} while in the S4/Sm_A the TUG-2 is mapped into a TUG-3 which is then mapped into a VC-4. Non-SDH client signals are mapped into a VC-3 {STS-1 SPE}.

4.2.8   The Path Protection Sub-layer SnP

Path protection is described in Chapter 7. Figure 4-9 shows the adaptation function used in the SnP sub-layer. In the sub-layer created for path protection

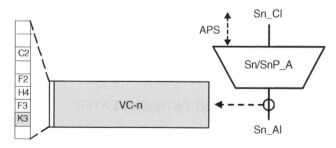

Figure 4-9. Sn Path protection adaptation function

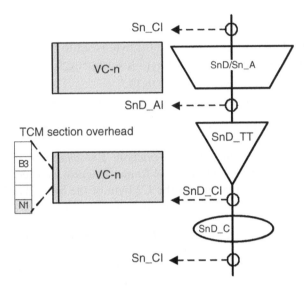

Figure 4-10. TCM sub-layer model

the adaptation function inserts and extracts the APS information from the K3[1 ... 4] bits in the Sn-POH. Path protection is supported only in higher order Sn layers, i.e., n = 3, 4.

4.2.9   The Tandem Connection Monitoring Sub-layer SnD

Tandem Connection Monitoring (TCM) is used to monitor the QoS of a trail section that may fall under the responsibility of a different operator. When TCM is enabled a sub-layer is created according to the functional modelling rules as depicted in Figure 4-10.

The TCM adaptation function SnD/Sn_A is an empty function; it is only present to complete the model. The TCM connection function has only one input and one output which are always connected. The TCM trail termination function uses only the *Network Operator* byte N1 (N2 in the Sm path layer) as the *TCM OverHead* (TCOH). Every change in the N1 (N2) byte affects the BIP value that is transferred in B3 byte (or V5[1,2] bits), this shall be compensated.

The TCOH byte is multiframed to provide more bits for the overhead information.

A detailed description can be found in [ITU-T Rec. G.707] annex D and E.

4.3   THE SDH LOWER ORDER PATH (Sm) LAYER

The path layer Sm, where "m" denotes the lower order *Virtual Container* (VC) type (m = 2, 12, 11), is the layer that provides the lower order connectivity in a network element.

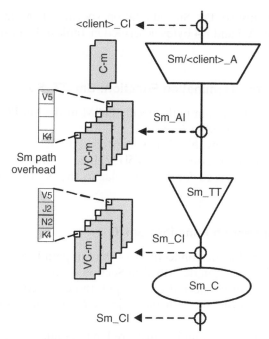

Figure 4-11. Sm network layer model

In SDH the S3 path layer is considered to be a lower order path layer. However, SONET considers the S3 path layer to be the higher order path layer and because the VC-3 {STS-1 SPE} frame structure is similar to the VC-4 frame structure, in this book the S3 path layer is described in Section 4.2.

The lower order path layer Sm serves as the transport layer for non-SDH tributary signal layers. It has the higher order path layer Sn, described in Section 4.2, as its server layer.

Figure 4-11 shows the generic Sm network layer model.

This model is applicable to both the SDH and SONET lower order Sm path layers (m = 2, 12, and 11). The C-m frame structure is described in Section 3.1.7 of chapter 3.

The Sm specific *Path OverHead* (Sm-POH) is used to transfer OAM information.

4.3.1   The Source Adaptation Function: Sm/<client>_A_So

This function maps and/or multiplexes the characteristic information of the <client> signal into the container C-m that is transported in the Sm path layer. Because there are several <client> signals, the specific adaptation functions Sm/<client>_A will be described separately after the generic description of the Sm atomic functions.

This function inserts the Sm adaptation specific OAM in the Sm-POH, i.e., the bits in the V5 and K4 bytes described in Table 4-3 and shown in Figure 4-11.

4.3.2   The Source Termination Function: Sm_TT_So

This function inserts the Sm path specific OAM in the Sm-POH, i.e., the V5, J2 and N2 bytes described in Table 4-4.

The API provisioned by the EMF is inserted in the J2 byte. The BIP-2 parity code is calculated and inserted in the V5[1, 2] bits. Normally the N2 byte is set to all-zero unless it is used for TCM similar to the N1 byte as described in Section 4.2.9.

4.3.3   The Connection Function: Sm_C

This function provides the connectivity of the Sm path layer and is similar to the Sn connection function described in Section 4.2.3.

If an output is not connected to an input, the connection function generates an Sm frame structure with the V5[5, 6, 7] bits set to the "000": "*path unequipped.*"

An SSF present at the input is transferred to the same output as the payload data.

In general path protection in the Sm path layer is not supported.

4.3.4   The Sink Termination Function: Sm_TT_Sk

This function extracts the Sm path specific OAM from the Sm-POH, i.e., the V5, J2, N2 and K4 bytes described in Table 4-5. It will use this information for checking the connectivity in the Sm layer, the Sm path status, and its performance.

If the API in the J2 byte is not equal to the ExTI, this will be reported to the EMF as TIM defect. If the C2 byte contains the value "000" the UNEQ defect is detected. If an SSF is received from the server layer, or the TIM or UNEQ defect is detected, it is transferred to the succeeding adaptation function as TSF and RDI is sent to the co-located Sm_TT_So function for insertion in the V5[8] bit. A received RDI is sent to the EMF to be used for fault handling. If a TIM or UNEQ defect is detected the output signal is replaced by an AIS (all-ones) signal. The BIP-2 received in the V5[1, 2] bits is compared to the calculated BIP-2. The number of detected bit-errors is used for near-end performance monitoring and sent as REI to the co-located Sm_TT_So function for insertion in the V5[3] bit. It is also used to detect the SD defect. This defect is forwarded to the following adaptation function as TSD. The received REI is used for local far-end performance monitoring. The value of the N2 byte is ignored unless it is used for TCM similar to the N1 byte described in Section 4.2.9.

4.3.5 The Sink Adaptation Function: Sm/<client>_A_Sk

Recovers or de-multiplexes the characteristic information of the <client> signal from the transported container C-m. The specifics for each possible adapted client signal are described in separate sections below.

The function will also extract the Sm adaptation specific OAM from the Sm-POH, i.e., the bits in bytes V5 and K4 described in Table 4-3.

The pointer process will report a LOP defect. If the value of the V5[5 ... 7] or K[1] bits does not match the mapping technology supported by this adaptation function a PLM defect is detected. If the H4 byte is used for multiframe alignment and cannot find alignment a LOM defect is detected. If a LOP or PLM or LOM defect is detected the output signal is replaced by an AIS (all-ones) signal and an SSF is sent to the client layer.

4.3.6 The Sm Specific Path Overhead: Sm-POH

The following two tables provide an overview and description of the Sm specific path overhead (Sm-POH) bytes.

Because each VC-m frame (repetition rate 125 µs) has only a single byte allocated for Sm-POH, bits 7 and 8 of the H4 byte located in the VC-n POH (see Table 4-8) are allocated to create a multiframe of 4 consecutive VC-m frames (500 µs) to provide 4 bytes that can be used for the VC-m POH as shown in Figure 4-12.

Table 4-3 contains the VC-m (m = 2, 12, 11) POH bytes dedicated to the mapping and/or multiplexing functionality of the Sm adaptation functions.

Table 4-4 contains the VC-m (m = 2, 12, 11) POH bytes dedicated to the path supervision functionality of the Sm trail termination functions.

Table 4-5 provides a description of the pointer bytes of a TU-m (m = 11, 12, 2) specified for the "floating" mapping functionality of the adaptation functions.

A TU-m frame (repetition rate 125 µs) has also only a single byte allocated for the pointer. To create a multiframe of 4 consecutive TU-m frames (500 µs) and provide 4 bytes that can be used for the TU-m pointer also the H[7, 8] bits are used. Figure 4-12 shows the relation between H4[7,8] and the TU-m pointer bytes V1, V3, V3, and V4.

4.4 THE SDH MULTIPLEX SECTION (MSn) TRANSPORT LAYER

The MSn layer, where n denotes the multiplex level 1, 4, 16, 64 or 256, has the S4 path layer as its client layer. In a SONET network the MSn layer normally has the S3 path layer as its client layer. The MSn layer has the RSn layer, described in Section 4.5, as its server layer. Figure 4-13 shows a functional model of the MS1 layer and its relation to the frame structures.

The multiplex section specific *OverHead* (MS-OH) is used to transfer OAM information. The STM-N frame structures are described in Chapter 3.

Sm-POH multiframe

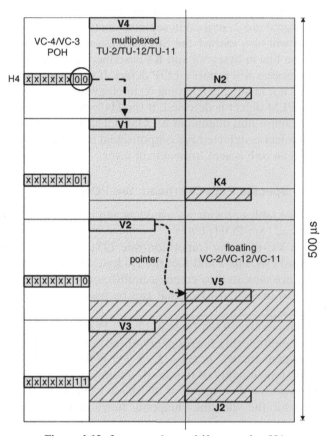

Figure 4-12. Lower-order multiframe using H4

The S4 path layer VC-n structure is mapped into the payload area of an *Administrative Unit* (AU-n) structure and the AU pointer (bytes H1, H2 and H3 described in Table 4-8) is used to enable the VC-n to "float" in the AU-n structure. The pointer process is described in Chapter 6. Next the AU-n structure is mapped into an *AU Group of level N* (AUG-N) structure. This mapping has a fixed phase relation.

Lower level AUG-N are column interleave multiplexed into higher level AUG-N structures. This multiplexing is fixed. Finally the AUG-N structure is mapped into an STM-N structure. This mapping is also fixed.

An MSn trail can be protected by creating a trail protection sub-layer according to the functional modeling rules. This is described in Section 7.3.1.1 of Chapter 7.

TABLE 4-3. VC-m (m = 2, 12, 11) adaptation overhead bytes

Byte	Description
V5	*VC-m path overhead* (POH) byte—provides the capability to check the error performance, send path status and payload mapping methodology. The bits assignment is as follows:

1	2	3	4	5	6	7	8
BIP-2		REI	RFI	Path Signal Label			RDI

Bits 5 ... 7: LO *Path Signal Label* (PSL)—provides eight binary values. Examples of *(binary) values for* V5[5 ... 7] are:

000   unequipped—open connection in the connection function
010   asynchronous mapping
100   byte synchronous mapping
101   extended signal label, an indication that K4[1] is multiframed and contains the signal label value
111   VC-AIS—indication of path AIS

A list of defined values is provided in Table 9-12 [ITU-T Rec. G.707]

Byte	Description
K4[1]	Extended signal label—32 consecutive K4[1] bits form a multiframe that contains the Extended signal label in case the signal label in V5[5 ... 7] has the value "101." Examples of Extended signal labels are: 0x09 ATM mapping 0x0A HDLC/PPP mapping 0x0D GFP mapping A list of defined values can be found in Table 9-13 [ITU-T Rec. G.707]
K4[2]	Virtual conCATenation (VCAT) multiframe—32 consecutive K4[2] bits form a VCAT multiframe, in phase with the K4[1] multiframe. Its use is described in Chapter 9.
K4[3,4]	Two APS signaling bits—allocated for automatic protection switching at VC-2/VC-12/VC-11 path levels. Currently for further study.
K4[5 ... 7]	Three undefined bits—reserved for future use, ignored at reception.

In SONET the lowest level MSn layer is the MS0. It is described in Section 4.4.7 because it has a mapping structure different from the generic MSn mapping. Because in SONET the higher order path layer is the S3 layer {STS-1 SPE} based structures are mapped into an AUG-1 structure before multiplexing the AUG-1 into higher order AUG-N structures. The MS1/S3_A function performs this mapping and is described in Section 4.4.8.

TABLE 4-4. VC-m (m = 2, 12, 11) termination overhead bytes

Byte	Description
V5	*VC-m path overhead* (POH) byte—provides the capability to check the error performance, send path status and payload mapping methodology. The bits assignment is as follows:

1	2	3	4	5	6	7	8
BIP-2		REI	RFI	Path Signal Label			RDI

- Bits 1 and 2: *Bit Interleaved Parity* (BIP-2)—is used for error performance monitoring. Bit 1 provides an even parity over all odd numbered bits in the previous VC-2/VC-12/VC-11 frame (including the path overhead bytes) and bit 2 provides similarly an even parity over all even numbered bits.
- Bit 3: Path *Remote Error Indication* (REI)—set to one if one or more errors were detected by the BIP-2 checker, and otherwise set to zero.
- Bit 4: byte synchronous Path *Remote Failure Indication* (RFI)—set to one if a failure is detected, otherwise it is set to zero. The use of this bit is defined only for VT-11 [VT-1.5].
- Bit 8: Path *Remote Defect Indication* (RDI)—set to one if a server signal failure or trail signal failure is detected by the termination sink function, otherwise it is set to zero.

Byte	Description
J2	Lower-Order path *Access Point Identifier* (API) byte—used to transfer a 16-byte frame as defined in clause 3 of [ITU-T Rec. G.831]. It is transmitted repetitively.
N2	*Network operator* byte—allocated to support a Lower-Order Tandem Connection Monitoring (LO-TCM) function.
K4[5 ... 7]	Three undefined bits—reserved for future use, ignored at reception.
K4[8]	Reserved for a path data link of 8 kbit/s. Currently not in use.

TABLE 4-5. TU-m (m = 2, 12, 11) pointer bytes

Byte	Description
V1, V2	*TU Pointer* bytes—specify the location of the VC frame that is floating in the TU frame, i.e., a pointer to the first byte (V5) of the VC in the TU payload area. The pointer mechanism is described in Chapter 6.
V3	*TU Pointer action* byte—this byte is used during frequency justification. Depending on the pointer action, the byte is used to adjust the fill input buffers. The byte only carries valid information in the event of negative justification, otherwise it is not defined.
V4	This byte is reserved for future use.

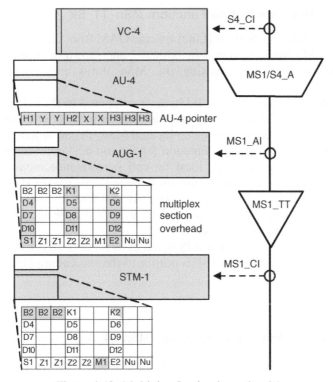

Figure 4-13. Multiplex Section layer, level 1

4.4.1 The Source Adaptation Function: MSn/S4_A_So

Maps the S4 characteristic information (a VC-4) into the container AUG-N that will be transported in the STM-N frame structure through the MSn network layer.

This function inserts the MSn adaptation specific overhead bytes in the MS-OH, i.e., the K1, K2, S1, D4 ... D12 and E2 bytes described in Table 4-6 and shown in Figure 4-13.

4.4.2 The Source Termination Function: MSn_TT_So

Inserts the Multiplex Section {Line} specific OAM in the MS-OH, i.e., the B2, K2 and M1 bytes described in Table 4-7 and shown in Figure 4-13.

The BIP-8×N code is calculated and inserted in the 3×N B2 bytes of an STM-N frame.

The co-located MSn_TT_Sk function provides the RDI and REI information that is inserted into the K2[6, 7, 8] bits and the M1 byte.

4.4.3 The Connection Function: MSn_C

The connection function consists of a single connection. It is generally not shown.

4.4.4   The Sink Termination Function: MSn_TT_Sk

Extracts the Multiplex Section {Line} specific OAM from the MS-OH, i.e., the B2, K2, and M1 bytes described in Table 4-7 and shown in Figure 4-13. This information is used for checking the MSn status and monitoring the performance.

The BIP-8×N received in the B2 bytes is compared to the calculated BIP-8×N. The number of detected bit-errors is counted for near-end performance monitoring and sent as REI to the co-located MSn_TT_So function for transfer to the remote MSn_TT_Sk function for far-end performance monitoring. The received REI is used for local far-end performance monitoring. The number of detected bit-errors is also used for the detection of the SD defect or the *Excessive* error (EXC) defect. The SD and EXC are sent to the next function as TSD resp. TSFProt and can be used by the protection switching mechanisms described in Chapter 7.

If the function detects an MS-AIS signal at its input from the RSn layer an AIS signal accompanied by a TSF is output to the succeeding MSn/Sn_A_Sk function.

4.4.5   The Sink Adaptation Function: MSn/S4_A_Sk

Recovers the S4 path layer characteristic information from the STM-N frame structure.

The pointer process will detect the LOP defect as well as the AU-AIS defect. When these defects are detected the function will output an AIS signal and an SSF to the client layer Sn_C function. The TSD, TSF, and TSFProt conditions received from the preceding MSn_TT_Sk function are forwarded to the Sn_C function as well.

The function will also extract the MSn adaptation specific OAM from the MS-OH, i.e., the K1, K2, S1, D4 ... D12 and E2 bytes described in Table 4-6.

4.4.6   The MSn Specific Section Overhead: MS-OH

The two following tables provide a description of the MSn (n = 0, 1, 4, 16, 64 and 256) specific MS-OH bytes. Table 4-6 contains the MS-OH bytes dedicated to the mapping and multiplexing functionality of the MSn adaptation functions.

Table 4-7 contains the MS-OH bytes dedicated to the path supervision functionality of the MSn trail termination functions.

Table 4.8 provides an overview and description of the pointer bytes of an AU-n and a TU-3 specified for the "floating" mapping functionality of the adaptation functions.

TABLE 4-6. MSn adaptation overhead bytes

Byte	Description
K1, K2[1 ... 5]	*Automatic Protection Switching* (APS) bytes—APS is a protection mechanism used to synchronize both ends of a protected entity. The APS protocol exchanges status information of the source bridge and the sink selector and it transfers switching requests based on detected defects and external commands.
D4 ... D12	Multiplex Section *Data Communications Channel* (MS DCC or DCCM) bytes—provide a 576 kbit/s message-based channel for the exchange of *Operations, Administration and Maintenance* (OAM) data between Multiplex Section terminating equipment.
D13 ... D156	Extended MS DCC (DCCMx) bytes—provide an extra 9,216 kbit/s message-based channel for the exchange of *Operations, Administration and Maintenance* (OAM) data between Multiplex Section terminating equipment. Only available in the STM-256.
S1	*Synchronization status message* (SSM) byte—Bits 5 to 8 of this byte are used to carry the synchronization status. The following synchronization levels are defined by the ITU-T (other values are reserved):
	0000 Quality unknown (for backwards compatibility) 0010 PRC [ITU-T Rec. G.811] 0100 SSU-A [ITU-T Rec. G.812] transit) 1000 SSU-B (G.812 local) 1011 SEC (G.813 Option 1) 1111 Do not use for synchronization. This level may be emulated by equipment failures and will be emulated by a Multiplex Section AIS signal.
E2	Multiplex Section *Engineering Order Wire* (EOW) byte—provides a 64 kbit/s channel between multiplex section terminating equipment. It can be used as a voice channel to be used by crafts persons and have its own interface on the equipment.

4.4.7 The SONET MS0 Layer

The MS0 layer, the lowest order SONET layer, has the S3 path layer as its client layer and the RS0 {section} layer as its server layer. Figure 4-14 shows the functional model of the MS0 layer and its relation to the frame structures.

At levels higher than the MS0 the {STS-1 SPE}, {STS-1-Xv SPE}, {STS-3c-Xv SPE}, and {STS-3nc SPE} are mapped and/or multiplexed into an AUG-N

TABLE 4-7. MSn termination overhead bytes

Byte	Description
B2	Multiplex Section *Error Detecting Code* (EDC) byte—in an STM-N frame structure 3 × N B2 bytes contain a 24 × N bit Bit Interleaved Parity code (BIP-24 × N), even parity. It is used to determine if a transmission error has occurred in a Multiplex Section. The BIP-24 × N is calculated over all bits in the MS Overhead and the payload of the previous STM-N frame before scrambling. The value is placed in the 3 × N B2 bytes of the MS-OH before scrambling.
K2[6 … 8]	*Multiplex Section Remote Defect Indication* (MS-RDI) bits, used to indicate to source side that the sink side has detected a Multiplex Section defect or is receiving MS-AIS.
M1	*Multiplex Section Remote Error Indication* (MS-REI) byte—conveys to the source side the count of the interleaved bit blocks that have been detected to be in error at sink side of the Multiplex Section.
M0	SONET: MS-REI byte used in the {STS-1} frame structure only. SDH: extra MS-REI byte. Used only in STM-64 and STM-256 frame structures to extend the counting capability of the M1 byte.
Nu	*National use*—bytes reserved for National use
Z1, Z2	*Growth* (for future use) byte—Only allocated in SONET, in SDH these are unused bytes.

TABLE 4-8. AU-n (n = 3, 4, 4-Xc) pointer bytes

Byte	Description
H1, H2	AU-n and TU-3 Pointer bytes—specify the location of the VC frame in the AU/TU frame, i.e., point to the first byte (J1) of the VC floating in the AU/TU payload area. The pointer process is described in Chapter 6.
H3	Pointer action byte—this byte is used during frequency justification. Depending on the pointer action, the byte is used to adjust the fill input buffers. The byte only carries valid information in the event of negative justification, otherwise it is not defined.
Y	In case the AUG-N is based on AU-4 each H1 byte is followed by two Y bytes. The Y byte is set to: "10010011."
X	In case the AUG-N is based on AU-4 each H2 byte is followed by two X bytes. The X byte is set to: "11111111."

for transport by an STM-N (N = 1, 4, 16, 64, and 256). The MS0 frame structure is described in Chapter 3.

The S3 path layer {STS-1 SPE} structure is mapped an AU-3 structure using the AU-3 pointer (bytes H1, H2 and H3 described in Table 4-8) to enable the

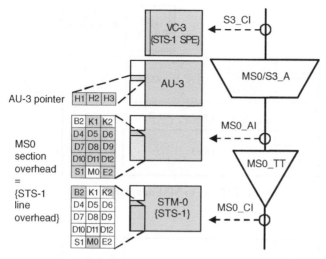

Figure 4-14. Multiplex Section layer, level 0

{STS-1 SPE} to "float" in the AU-3 structure. The pointer process is described in Chapter 6. Next the AU-3 structure is mapped into an STM-0 structure. The phase relation between these two structures is fixed.

The processes in the MS0/S3_A and the MS0_TT functions are the same as the processes in the MSn/Sn_A and MSn_TT functions described above. The available MS-OH bytes are shown in Figure 4-14 and described in Tables 4-6 and 4-7.

4.4.8   The S-3 to MS1 Adaptation Function

Normally (in SDH) a VC-3 is transported by multiplexing it into a VC-4 because the S4 is the higher order path layer. This is described in Section 4.2.7 and shown in Figure 4.4.

In SONET the S3 path layer is the higher order layer. Here the VC-3 is mapped into an AU-3 structure that is multiplexed into an AUG-1 structure for further transport.

The {STS-1 SPE} is mapped into an AU-3 structure using the AU-3 pointer (bytes H1, H2, H3) to enable the {STS-1 SPE} to "float" in the AU-3. Three AU-3 are column interleave multiplexed into a AUG-1. Figure 4-15 shows the mapping and multiplexing of the S3 client signal into the AUG-1.

4.5   THE SDH REGENERATOR SECTION (RSn) LAYER

The RSn layer, where n denotes the multiplex level 1, 4, 16, 64, or 256, has the MSn layer as its client layer and the optical OSn layer (Section 4.6) or

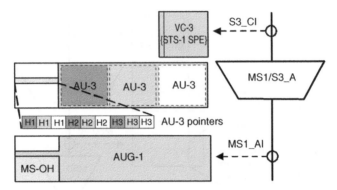

Figure 4-15. S3 to MS1 adaptation via AU-3

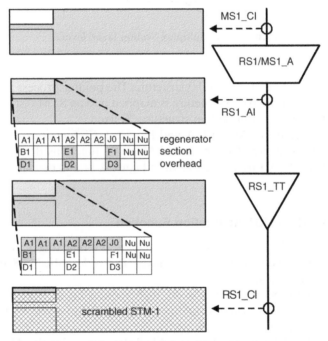

Figure 4-16. Regenerator Section layer, level 1

electrical ESn layer (Section 4.7) as its server layer. In an OTN network it will have the ODU layer as its server layer (see Section 4.10).

Figure 4-16 shows the functional model of the RS1 layer and its relation to the frame structures. The STM-N frame structures are described in Chapter 3.

No mapping or multiplexing is performed in this layer of the network. It has dedicated *Regenerator Section OverHead* (RS-OH) to provide the capability to check the connectivity, monitor the performance, and communicate with the regenerators located between two network elements.

Forward Error Correction (FEC) can be used to correct bit errors that occur due to the extended range of the optical signals. The FEC capability is provided by a dedicated adaptation function as described in Section 4.5.8.

4.5.1   The Source Adaptation Function: RSn/MSn_A_So

This function inserts RSn adaptation specific OAM in the RS-OH area of the STM-N frame structure, i.e., the D1 … D3, E1 and F1 bytes described in Table 4-9 and shown in Figure 4-16.

4.5.2   The Source Termination Function: RSn_TT_So

This function inserts the Regenerator Section {Section} specific OAM in the RS-OH area of the STM-N frame structure, i.e., the A1, A2, J0, and B1 bytes described in Table 4-10 and shown in Figure 4-16.

The API value provisioned by the EMF is inserted in the J0 byte. The BIP-8 parity code is calculated over the previous scrambled STM-N frame and inserted in the B1 byte of the current STM-N frame before scrambling.

Finally the STM-N frame is scrambled. The scrambling does not include the first row of the RS-OH: bytes [1, 1 … 9N]. Scrambling ensures that there is a balance between the number of "0" bits and "1" bits and avoids long strings with bits having the same value.

4.5.3   The Connection Function: RSn_C

The connection function consists of a single connection. It is generally not shown.

4.5.4   The Sink Termination Function: RSn_TT_Sk

Uses the A1 and A2 bytes to recover the STM-N frame structure from the bit stream received from the physical layer. It descrambles the STM-N frame and extracts the Regenerator Section {Section} specific OAM from the RS-OH, i.e., the J0 and B1 bytes described in Table 4-10 and shown in Figure 4-16. This information is used for checking the connectivity in the RSn layer, the RSn section status, and its performance.

If the API in the J0 byte is not equal to the ExTI, this will be reported to the EMF as TIM defect. If TIM is detected, and TIM consequent actions are enabled, the output signal is replaced by an MS-AIS signal and TSF is sent to the succeeding RSn/MSn_A_Sk function. The MS-AIS and TSF will also be output if an SSF condition is received from the server layer. The BIP-8 received in the B1 byte is compared to the calculated BIP-8. The number of detected

TABLE 4-9. RSn adaptation overhead bytes

Byte	Description
E1	RS Engineering Order Wire (RS-EOW) byte—provides a 64 kbit/s channel between regenerator section elements. It can be used as a voice channel to be used by crafts persons.
F1	RS User Channel byte—used for user communication between regenerator section elements.
D1 ... D3	Regenerator Section Data Communications Channel (RS DCC or DCCR) bytes—provide a 192 kbit/s message-based channel for the exchange of Operations, Administration and Maintenance (OAM) data between regenerator section terminating equipment.
P1	*Forward Error Correction* (FEC) bytes—can contain the FEC code. The FEC code is calculated separately for each row in the STM-N frame structure. The P1 bytes are located in the unused bytes of the MS-OH and RS-OH. Used only for RS16, RS64 and RS256.
Q1[7,8]	*FEC Status Indicator* (FSI) bits, used to indicate the status of the encoder. The value "01" indicates that the encoder is on.

bit-errors is counted for near-end performance monitoring, and reported to the EMF.

4.5.5 The Sink Adaptation Function: RSn/MSn_A_Sk

Extracts the RSn adaptation specific OAM from the RS-OH area in the STM-N frame structure, i.e., the D1 ... D3, E1 and F1 bytes described in Table 4-9 and shown in Figure 4-16. A TSF received from the RSn_TT_Sk function is transferred as an SSF to the client MSn section layer.

4.5.6 The RSn Specific Section Overhead: RS-OH

The two following tables provide an overview and description of the RSn (n = 0, 1, 4, 16, 64 and 256) specific overhead (RS-OH) bytes.

Table 4-9 contains the RS-OH bytes dedicated to the mapping and multiplexing functionality of the RSn adaptation functions.

Table 4-10 contains the RS-OH bytes dedicated to the path supervision functionality of the RSn trail termination functions.

4.5.7 The SONET RS0 Layer

The RS0 layer, the lowest order SONET {section} layer, has the MS0 {link} layer as its client layer and the OS0 layer or ES0 layer as its server layer.

Figure 4-17 shows the functional model of the RS0 layer and the related frame structures. The processes in the RS0/MS0_A and RS0_TT functions are

TABLE 4-10. RSn termination overhead bytes

Byte	Description
A1, A2	Framing bytes—two bytes indicating the beginning of the STM-N frame. The A1, A2 bytes remain unscrambled. A1 has the binary value 11110110, and A2 has the binary value 00101000.
	The frame alignment word of an STM-N frame consists of $(3 \times N)$ A1 bytes followed by $(3 \times N)$ A2 bytes, except the STM-256 frame that has 64 A1 bytes in $[1, 705 \dots 768]$ adjacent to 64 A2 bytes in $[1, 769 \dots 832]$.
J0	Regenerator *Section Access Point Identifier* (SAPI) byte—used to transfer repetitively a 16-byte frame as defined in clause 3 of [ITU-T Rec. G.831].
B1	Regeneration Section *Error Detecting Code* (EDC) byte—contains an 8 bit Bit Interleaved Parity code (BIP-8), even parity, it is used to check for transmission errors in a regenerator section. Its value is calculated over all bits of an STM-N frame after it has been scrambled and then placed in the B1 byte of the next STM-N frame before it is scrambled.
Nu	*National use*—bytes reserved for National use.
Z0	*International use*—bytes reserved for International use, the first $(N-1)$ bytes after the J0 byte for $N = 4, 16, 64$ and 256.

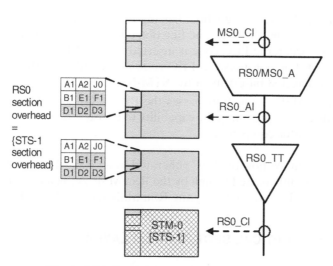

Figure 4-17. Regenerator Section layer, level 0

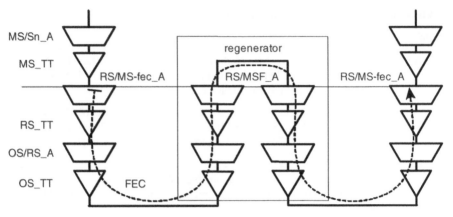

Figure 4-18. FEC transparent transport

the same as the processes in the RSn/MSn_A and RSn_TT functions described above.

4.5.8 The FEC Capable RSn to MSn Adaptation Functions

If the FEC coding is enabled the FEC code is generated per row of the STM-N (N ≥ 16) frame structure, the RS-OH area is excluded from the FEC calculation.

There are two adaptation functions defined in [ITU-T Rec. G.783] that support FEC:

- The RSn/MSn-fec_A function: has the capability to generate and insert FEC code in the P1 bytes. The function also corrects the BIP code in the B2 bytes when the FEC code is inserted in the P1 bytes. The RSn/MSn-fec_A_Sk function uses the FEC code recovered from the P1 bytes to correct detected bit-errors in the STM-N frame.
- The RSn/MSF_A function: passes the P1 and Q1 bytes transparently in both directions providing the capability to transfer the FEC transparently in regenerator equipment.

Figure 4-18 shows how the FEC generated by an RSn/MSn-fec_A_So function is transparently passed trough by the RSn/MSF_A_Sk and RSn/MSF_A_So functions and is terminated by the RSn/MSN-fec_A_Sk function.

4.6 THE SDH OPTICAL SECTION (OSn) LAYER

The OSn layer, where n denotes the multiplex level 1, 4, 16, 64, or 256, is one of the two possible physical layers in a network. This layer has the RSn

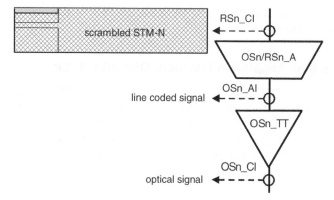

Figure 4-19. Optical Section layer, level n

layer as its client layer. It has an optical fiber as its server layer as shown in Figure 4-19.

By selecting the appropriate optics, the optical signal can be used intra-station, and for short haul, long haul, very long haul, and ultra long haul inter-station applications.

4.6.1 The Source Adaptation Function: OSn/RSn_A_So

The scrambled STM-N frame structure received from the preceding RSn_TT_ So function is converted into the line coded signal as defined in [ITU-T Rec. G.957] or [ITU-T Rec. G.691]. This function limits the output jitter to the values defined in [ITU-T Rec. G.825].

4.6.2 The Source Termination Function: OSn_TT_So

This function generates an optical STM-N signal that meets the optical characteristics defined in [ITU-T Rec. G.957] or [ITU-T Rec. G.691] for transmission over the optical medium. A *Loss of Signal* (LOS) detected by the co-located OSn_TT_Sk function can be used for laser safety as defined in [ITU-T Rec. G.664].

4.6.3 The Connection Function

The connection function consists of a single connection. It is generally not shown.

4.6.4 The Sink Termination Function: OSn_TT_Sk

This function receives the STM-N line signal transmitted over the optical medium and converts it to the STM-N binary signal. It will detect the LOS defect. LOS can be sent to the co-located OSn_TT_So function for laser safety, will be reported to the EMF and signaled, as TSF to the succeeding OSn/RSn_A_Sk function.

4.6.5 The Sink Adaptation Function: OSn/RSn_A_Sk

Processes the OSn_AI signal to recover the *data signal* (D) and the associated *clock signal* (CK). The framing process provides the *frame-start signal* (FS) and will detect the *Loss of Frame* (LOF) defect. LOF will be reported to the EMF and signaled as SSF to the succeeding RSn_TT_Sk function. A TSF condition received from the preceding OSn_TT_Sk function is forwarded as SSF to the RSn_TT_Sk function.

4.7 THE SDH ELECTRICAL SECTION (ES1) LAYER

The ES1 layer is the second of the two physical layers in an SDH and SONET network. In SDH only the first multiplex level has an electrical interface. SONET has two levels of multiplexing with an electrical interface, the *Electrical Carrier Level 1* (EC-1) and the *Electrical Carrier level 3* (EC-3).

This layer has the RS1 layer as its client layer. It has an electrical cable as its server layer as shown in Figure 4-20. Generally the application of the ES1 is intra-station.

The model in Figure 4-20 can also be used for the SONET {EC-1} and {EC-3} signals.

4.7.1 The Source Adaptation Function: ES1/RS1_A_So

The scrambled STM-1 frame structure received from the preceding RS1_TT_So function is converted into a *Coded Mark Inversion* (CMI) encoded STM-1

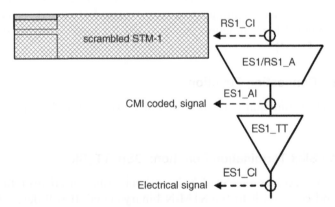

Figure 4-20. Electrical Section layer, level 1

signal. In SONET the {STS-1} frame structure is converted into a *Bipolar with Three-Zero Substitution* (B3ZS) encoded {EC-1} signal and an {STS-3} is encoded into a CMI encoded {EC-3} signal.

4.7.2   The Source Termination Function: ES1_TT_So

Generates an STM-1 line signal that meets the electrical characteristics defined in [ITU-T Rec. G.703]; *Pulse shape, Peak-to-peak voltage, Rise time*, and *Output return loss*. The characteristics of the SONET {EC-1} and {EC-3} signals are defined in [ANSI T1.102].

4.7.3   The Connection Function

The connection function consists of a single connection. It is generally not shown.

4.7.4   The Sink Termination Function: ES1_TT_Sk

This function receives the STM-1 line signal transmitted over the electrical cables and converts it to the STM-1 binary signal. It will detect the LOS defect. LOS will be reported to the EMF and signaled as TSF to the succeeding ES1/RS1_A_Sk function. In SONET the {STS-1} is recovered from the {EC-1} signal and the {STS-3} recovered from the {EC-3} signal.

4.7.5   The Sink Adaptation Function: ES1/RS1_A_Sk

This function processes the ES1_AI signal to recover the *data signal* (D) and the associated *clock signal* (CK). The framing process provides the *frame-start signal* (FS) and will detect the LOF defect; the LOF will be reported to the EMF and signaled as SSF to the succeeding RS1_TT_Sk function. A TSF condition received from the preceding ES1_TT_Sk function is forwarded as SSF to the RS1_TT_Sk function.

4.8   THE LAYERS BELOW THE SDH OSn AND ES1 LAYER

The optical fibers that carry the OSn signals or the electrical cables that carry the ES1 signal can be bundled (or multiplexed) again and run through a duct. This duct can be considered as a layer as well and modeled according to the functional modeling.

4.9   THE OPTICAL TRANSPORT NETWORK OTN

Figure 4-21 shows the functional model of the OTN network and the related frame structures. This particular model is used to show the different layers that

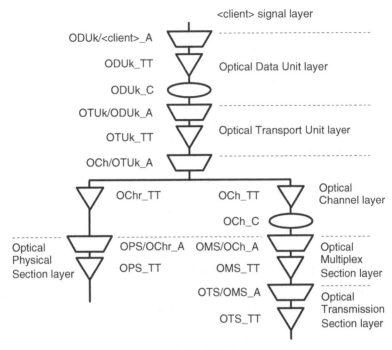

Figure 4-21. OTN layers

are identified in the OTN network; it does not represent an equipment model or a network model.

The descriptions of each network layer that follow are ordered according to Figure 4-21 from the top to the bottom. Within each network layer, descriptions of the three atomic functions are provided. This is a combination of the specifications described in [ITU-T Rec. G.709] and [ITU-T Rec. G.798].

4.10   THE OTN OPTICAL CHANNEL DATA UNIT LAYER (ODUk)

The ODUk layer, where k denotes the level of multiplexing 1, 2, 3 (and soon 4) is used to transport CBR signals or packet based signals received from the client layer. The ODUk layer can have either a higher order ODUk layer or an OTUk layer of the same or a higher level as its server layer. Figure 4-22 shows the atomic functions in the ODUk layer. Client signals are mapped in the OPUk payload area for further transport. Frequency justification is performed by providing single octet positive and negative justification opportunities per frame. The OTN has no pointer processing like SDH.

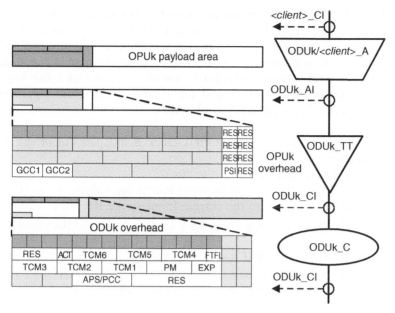

Figure 4-22. ODUk layer

4.10.1   The Source Adaptation Function: ODUk/<client>_A_So

Maps and/or multiplexes the characteristic information of the <client> signals into the payload area of the *Optical channel Payload Unit of level k* (OPUk) that is transported in the ODUk layer. Currently (2008) the following client signals are supported:

1. Constant Bit-Rate signals: CBR2G5, CBR10G, CBR40G
2. SDH STM-N (N = 16, 64, 256) frame structures
3. ATM cells
4. Packet based signals using the GFP mapping methodology
5. Lower order ODUk frame structures
6. Test signals: e.g., PRBS, NULL

In general this function inserts the adaptation specific OAM in the OPU-OH, i.e., the *Payload Type* (PT) and *Justification OverHead* (JOH). The values used in the fields are specified at the specific adaptation function.

If one or both of the *Generic Communications Channel* (GCC) is enabled the GCC1 and GCC2 fields in the ODU-OH are used for communication between network management and control systems.

The following sections contain client signal specific adaptation descriptions.

4.10.1.1 The CBR to ODUk Adaptation: ODUk/CBRx_A

In the ODUk/CBRx_A (x = 2G5, 10G, 40G) function constant bitrate signals are mapped into the payload area of an OPUk. Both a-synchronous and bit-synchronous mapping is supported. Figure 4-23 shows this specific adaptation function.

The OPUk JOH contains fields for *Justification Control* (JC), the *Negative Justification Opportunity* (NJO), and the *Positive Justification Opportunity* (PJO) as described in Section 6.3 of Chapter 6.

- A-synchronous mapping, the OPUk clock is independent of the client signal clock:
  The PT byte in the PSI is set to "0x02."
  The ODUk/CBRx_A_So function inserts the values of the JC, NJO and PJO fields related to the current frame in the JOH of this frame.
- Bit-synchronous mapping, the OPUk clock is locked to the client signal clock:
  The PT byte in the PSI is set to "0x03."
  Because no justification is required the values in the JOH are fixed: JC and NJO are set to "0x00" and PJO contains client data.

The ODUk/CBRx_A_Sk function is common for both mappings, it extracts the JC information and acts accordingly. It will detect and report *PayLoad Mismatch* (PLM) if the received and *accepted PT* (AcPT) is not equal to "0x02" or "0x03." It will consequently output an AIS and SSF signal and provide the received AcPT to the EMF. The AIS and SSF will also be output if a TSF is received from the preceding ODUk_TT_Sk function, see also Section 4.10.4.

4.10.1.2 The SDH to OTN Adaptation: ODUk/RSn_A

Typical 2.5, 10, and 40 Gbit/s CBR signals are the SDH rates STM-16, STM-64, and STM-256. The adaptation of the SDH RSn layer to the OTN ODUk path layer is performed by the ODUk/RSn_A function with k = 1, 2, 3 and n = $4^{(k+1)}$.

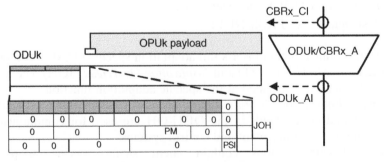

Figure 4-23. ODUk/CBRx_A

The ODUk/RSn_A function is connected to the RSn_TT function as shown in Figure 4-24. The RSn termination function is described in Section 4.5.

The mapping methodology used in the ODUk/RSn_A function is the same as used in the ODUk/CBRx_A function described in the previous Section 4.10.1.1.

4.10.1.3 The ATM Virtual Path to ODUk Adaptation: ODUk/VP_A

Adapts the ODUk path layer to the ATM *Virtual Path* (VP) layer. The OPUk payload area is filled with ATM cells received at the VP connection points. The total ATM cell stream bitrate shall match exactly the capacity of the OPUk payload structure.

The payload area of the ATM cells is scrambled before mapping into the OPUk. This is shown in Figure 4-25. The ATM octet structure is aligned with the ODUk octet structure.

The ODUk/VP_A_So function inserts at each input the *VP Identifier* (VPI) in each ATM cell. The *Header Error Control* (HEC) code is calculated and

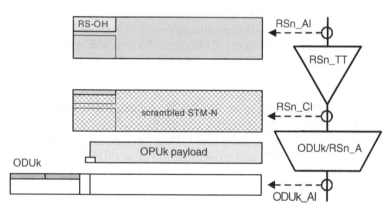

Figure 4-24. ODUk/RSn_A and RSn_TT functions

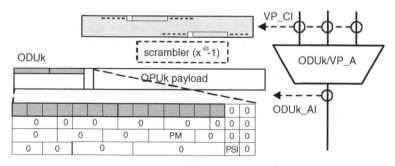

Figure 4-25. ODUk/VP_A

inserted in the HEC field of each ATM cell. In the OPU-OH the PT byte is set to "0x04."

The ODUk/VP_A_Sk function detects and reports PLM if the AcPT is not equal to "0x04." It will consequently output AIS and SSF signals and provide the AcPT to the EMF. The AIS and SSF will also be output if a TSF is received from the preceding ODUk_TT_Sk function (see also Section 4.10.4).

4.10.1.4 The PTN to OTN Adaptation This function adapts packet technology–based client layer signals for transport in the OTN layers. The client signal PDUs are mapped into the OPUk payload area using the GFP mapping methodology as described in Section 8.3 of Chapter 8. Figure 4-26 shows the functional model of this adaptation function with example <client> signals. That is, the Ethernet MAC layer (ETH) [ITU-T Rec. G.8021] and the MPLS-TP (MT) layer [ITU-T Rec. G.8121] as indicated by the ETH_TT and MT_TT functions.

More client layer mappings can be found in [ITU-T Rec. G.7041].

The ODUk/<client>_A_So function sets the PT byte of the OPU-OH to "0x05".

The ODUk/<client>_A_Sk function will detect and report PLM if the AcPT is not equal to "0x05." It will consequently output AIS and SSF signals and provide the received AcPT to the EMF. The AIS and SSF will also be output if a TSF is received from the preceding ODUk_TT_Sk function (see also Section 4.10.4).

4.10.1.5 The ODU[i]j (j < k) to ODUk Adaptation: ODUk/ODU[i]j_A Performs the multiplexing of ODUj structures into an ODUk structure. Up to i ODUj can be multiplexed into the OPUk payload area of an ODUk, i.e., four ODU1 into an ODU2, and a combination of m ODU2 and n ODU1 into an ODU3 with $(4m + n) \leq 16$.

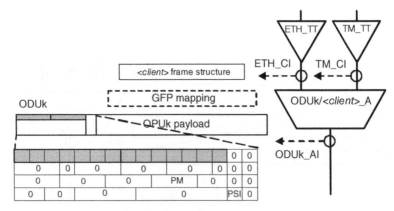

Figure 4-26. ODUk/<client>_A

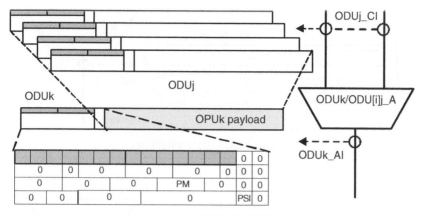

Figure 4-27. ODUk/ODU[i]j_A

Figure 4-27 shows this adaptation function.

The ODUk/ODU[i]j_A_So function inserts frame/multiframe alignment overhead in each of the individual ODUj, i.e., writes the FAS information in [1, 1 ... 7]. Frequency justification is performed per individual ODUj by using elastic buffers and the JOH of each ODUj as described in Section 6.3.2 of Chapter 6. No justification is needed for the ODUk; hence its JOH is set to all zero. The used mapping scheme is indicated in the *Multiplex Structure Identifier* (MSI) field described in Table 4-11. The function also sets the PT byte in the OPU-OH to "0x20."

The ODUk/ODU[i]j_A_Sk function will detect and report a PLM if the AcPT does not match the value "0x20." It will also detect and report an *MSI Mismatch* (MSIM) if the received and *accepted MSI* values (AcMSI) do not match the expected MSI values. In both cases the function will consequently output AIS and SSF signals at each individual ODUj and provide the AcPT and AcMSI to the EMF.

The function will use the MSI information to recover the individual tributary ODUj structures. It will perform frame/multiframe alignment on each tributary signal. If the alignment fails for an ODUj a *Loss of Frame/Multiframe* (LOF/LOM) is detected and reported. An AIS and SSF signal is output for this ODUj. The AIS and SSF will also be output if a TSF is received, and an SSD is output if a TSD is received from the preceding ODUk_TT_Sk function, as described in Section 4.10.4.

4.10.1.6 Test Signals for ODUk Path   Two types of test signals are supported, a *Pseudo Random Bit Sequence* (PRBS) signal and an all ZERO (NULL) signal.

4.10.1.6.1 The PRBS Signal to ODUk Adaptation Function ODUk/PRBS_A
The ODUk/PRBS_A function is used for testing purposes. Figure 4-28

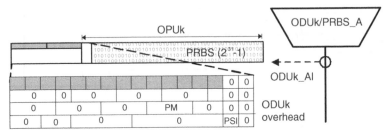

Figure 4-28. ODUk/PRBS_A

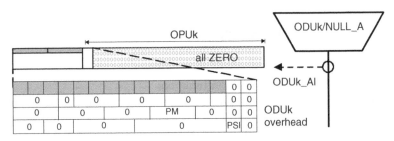

Figure 4-29. ODUk/NULL_A

shows the functional model and the frame structure of the adapted information

The ODUk/PRBS_A_So function generates a self-synchronizing $(2^{31}-1)$ PRBS signal and inserts it into the OPUk payload area. In the OPU-OH the PT byte is set to "0xFE." The ODUk/PRBS_A_Sk function will detect and report a PLM if the AcPT is not equal to "0xFE." It will consequently output an AIS and SSF signal and provide the AcPT to the EMF. The function will also extract the PRBS signal from the OPUk payload area. If it is unable to lock onto the PRBS signal a *Loss of Sequence Synchronisation* defect (dLSS) is detected. Bit errors in the PRBS signal, i.e., *Test Sequence Errors* (TSE) are counted per second and are used for performance monitoring.

4.10.1.6.2 The NULL Signal to ODUk Adaptation Function ODUk/NULL_A
The ODUk/NULL_A function is used for testing purposes. Figure 4-29 shows the functional model and the frame structure of the adapted information.

The ODUk/NULL_A_So function inserts an "*all ZERO*" signal into the OPUk payload area. In the OPU-OH the PT byte is set to "0xFD."

The ODUk/NULL_A_Sk function will detect and report a PLM if the AcPT is not equal to "0xFD." It will consequently output an AIS and SSF signal and provide the received value as AcPT to the EMF.

4.10.2   The Source Termination Function: ODUk_TT_So

This function inserts ODUk path specific overhead bytes in the PM field of the ODU-OH. A detailed description of the PM field is provided in Table 4-12.

The API provisioned by the EMF is inserted in the TTI byte. The STAT bits are set to "110": *normal path.*" The backward indicator bits BDI and BEI are set according to the information received from the co-located ODUk_TT_Sk function. And finally the BIP-8 code is calculated and inserted in the PM field. The ACT and TCMi fields are set to all-zero, they are used for *Tandem Connection Monitoring* (TCM) as described in Section 4.10.6.

4.10.3   The Connection Function: ODUk_C

This function provides the connectivity of the ODUk path layer. It has the same functionality as the path connection function Sn_C in SDH described in Section 4.2.3.

If an output of the ODUk_C function is not connected to an input an ODUk frame is generated with the STAT bits in the PM field set to "110": "*open connection.*"

The ODUk_C function can provide path protection in the ODUk layer as described in detail in Chapter 7. In this case the APS/PCC field in the ODU_OH is used to transfer the protection switching information. Protection switching can be initiated by the SSF and SSD signals received from the server layer.

An SSF that is present at the input is transferred to the same output as the payload data.

4.10.4   The Sink Termination Function: ODUk_TT_Sk

Extracts ODUk path specific OAM from the PM field in the ODU-OH which is used for checking the connectivity in the ODUk layer, the ODUk path status and performance monitoring. A detailed description of the PM field is provided in Table 4-12.

If the API in the TTI byte is not equal to the ExTI, this will be reported to the EMF as TIM defect. The STAT bits are checked to indicate "*normal path,*" if the STAT bits indicate AIS, LCK, or OCI the associated defect will be reported. If one of the above defects is detected the output signal is replaced by an AIS signal, the fault status is reported to the succeeding adaptation function as TSF and BDI is sent to the co-located ODUk_TT_So function for transfer to the far-end. The received BDI is reported to the EMF for fault localisation.

The received BIP-8 code is compared to the calculated BIP-8. The number of detected block-errors is counted for near-end performance monitoring and sent to the co-located ODUk_TT_So function for transfer to the far end in the BEI bits, it is also used for the detection of an SD defect. If SD is detected a TSD is sent to the succeeding adaptation function. The block-error count received via the BEI bits is used for local far-end performance monitoring.

The ACT and TCMi fields are ignored unless they are used for *Tandem Connection Monitoring* (TCM) as described in Section 4.10.6.

4.10.5 The Sink Adaptation Function: ODUk/<client>_A_Sk

This function recovers or de-multiplexes the characteristic information of the <client> signal from the transported OPUk structure. The specifics for each possible adapted client signal are described in separate subsections of Section 4.10.1 above.

The adaptation specific OAM of the OPUk, i.e., the PT and JOH, is extracted. The values used in the fields are specified at the specific adaptation function.

If one or both of the GCC is enabled the GCC1 and GCC2 fields in the ODU-OH are used for communication between network management and control systems.

An AIS and SSF signal will be output if a TSF is received, and an SSD is output if a TSD is received from the preceding ODUk_TT_Sk function, as described in Section 4.10.4.

4.10.6 Tandem Connection Monitoring (TCM) in the OTN

TCM is used to monitor the QoS of a trail section that may fall under the responsibility of a different operator. When TCM is enabled a sublayer is created according to the functional modeling rules as shown in Figure 4-30. The TCM specific overhead, i.e., the ACT and the TCMi fields in the ODU-OH are described in Table 4-12. The *TCM Control* function (ODUkT_TCMC) coordinates the processing of the TCM overhead.

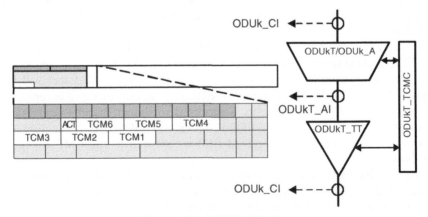

Figure 4-30. ODUk TCM layer

Subclause 14.5.1 in [ITU-T Rec. G.789] provides a detailed description of OTN TCM.

4.10.7   The ODUk Specific Adaptation Overhead: OPU-OH

The payload area of an ODUk frame structure is the OPUk and the OPUk frame structure consists of the payload area and the adaptation specific overhead OPU-OH.

Table 4.11 provides a description of the OPUk (k = 1, 2, 3) specific adaptation overhead fields. The overhead is located in [1 ... 4, 15 16] of the ODUk frame structure.

4.10.8   The ODUk Specific Path Overhead: ODU-OH

Table 4.12 provides a description of the ODUk (k = 1, 2, 3) path specific overhead fields. The overhead is located in [2 ... 4, 1 ... 14] of the ODUk frame structure.

TABLE 4-11. OPUk path overhead fields: OPU-OH

Field	Description
PSI	*Payload Structure Identification* field—located in byte [4, 15] of the ODUk structure. 256 consecutive bytes, aligned with the ODUk multiframe, constitute the PSI field. PSI[0] is used for the PT, PSI[1 ... 256] are used by mapping and concatenation specific functions.
PT	*Payload Type* field—one byte located in PSI[0], it is used to identify the mapping or multiplexing methodology used in the OPUk. It is similar to the SDH signal label (C2) as described in Table 4-1. Some examples: 0x02 a-synchronous CBR 0x03 bit-synchronous CBR 0x04 ATM mapping 0x05 GFP mapping 0x20 ODU structured 0xFD NULL test signal 0xFE PRBS test signal
MSI	*Multiplex Structure Identifier* field—16 bytes located in PSI[2 ... 17], it is used to indicate the ODUk type of a tributary port and the timeslot in the OPUk used by a tributary port. The indication for timeslot i is located in PSI[i + 1] (i = 1 ... 16), the two MSB indicate the ODUk type (value = k − 1) and the 6 LSB indicate tributary port number (value = 0 ... 15).
RES	*Reserved* field—the bytes in [1 ... 3, 15] and [1 ... 4, 16] are reserved for use by specific mapping and concatenation functions.

TABLE 4-12. ODUk path overhead fields ODU-OH

Field	Description
ACT	TCM *Activation* (ACT) field—one byte used as activation/deactivation channel for TCM (its use is for further study)
PM	*Path Monitoring* (PM) field—consists of three bytes:
	TTI *Trail Trace Identifier.* A 64-byte string containing: —a 16-byte source *Access Point Identifier* (API), —a 16-byte destination API and —a 32-byte operator specific field.
	BIP-8 *8-Bit Interleaved Parity* code calculated over the OPUk frame i (column 15—3624) and inserted in the BIP-8 byte in frame i + 2.

	status bits [1 … 4]—*Backward Error Indication* (BEI) block count
	bit [5] —*Backward Defect Indication* (BDI) if set to "1"
	bits [6 … 8]—Path *Status* (STAT) indication:
	  * normal path "001"
	  * locked—ODUk-LCK "101"
	  * open connection—ODUk-OCI "110"
	  * alarm indication—ODUk-AIS "111"

TCMi	*Tandem Connection Monitor level i* (i = 1 … 6) fields—each field consists of three bytes:
	TTIi the same as PM TTI
	BIP-8 the same as PM BIP-8

	status bits [1 … 4]—BEI block count, the value "1011" indicates a *Backward Incoming Alignment Error* (BIAE).
	bit [5] —*Backward Defect Indication* (BDI) if set to "1"
	bits [6 … 8]—additional Path *Status* (STAT) indication:
	  * in use no IAE "001" (= normal path)
	  * in use with IAE "010"

EXP	*Experimental* field—two bytes for experimental use.
APS/PCC	*Automatic Protection Switching* and *Protection Communication Channel* field—consist of four bytes. Up to eight levels of connection monitoring are supported: 1 for the ODUk path, 6 for the TCM levels and 1 for the OTUk section level.
FTFL	*Fault Type and Fault Localisation* communication channel—one byte allocated to transport a 256 byte FTFL message (for further study).
GCC1 and GCC2	*Generic Communications Channel* (GCC) fields—two times two bytes that provide two clear channels between any two network elements that have access to the ODU-OH.
RES	*Reserved* (RES) fields—nine bytes, these are for future use.

4.11   THE OTN OPTICAL CHANNEL TRANSPORT UNIT LAYER (OTUk)

The OTUk layer, where k denotes the level of multiplexing 1, 2, 3 (and 4) has the ODUk path layer of the same level as its client layer, it has the OCh layer as its server layer. Figure 4-31 shows how the layers are connected and how the signals are processed by each atomic function in the OTUk layer. The ODUk client signal is mapped frame synchronously in the OPUk server signal.

4.11.1   The Source Adaptation Function: OTUk/ODUk_A_So

This function creates the OTUk signal and maps the ODUk signal frame synchronous into this signal. The OTUk clock is generated by multiplying the received ODUk clock by a factor 255/239. If the ODUk frame does not start at the expected frame start position an IAE defect is detected and reported to the EMF and the next function. If the function is provisioned "locked" the input ODUk_CI signal is replaced by a LCK signal. If the GCC is enabled the GCC0 field in the OTU-OH is used for this channel. The channel can be used by the network management and control systems.

4.11.2   The Source Termination Function: OTUk_TT_So

This function inserts the OTUk section specific OAM in the SM field of the OTU-OH. A detailed description of the SM field is provided in Table 4-13.

The API value provisioned by the EMF is inserted in the TTI bytes of the SM field. The bits BDI, BEI, and BIAE are set in the SM field according to

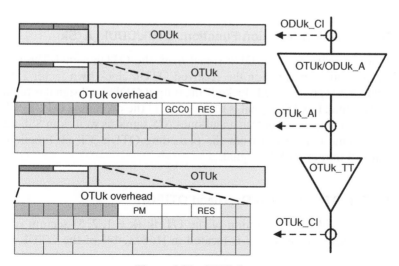

Figure 4-31. OTUk layer

the information received from the co-located OTUk_TT_Sk function. If the preceding OTUk/ODUk_A_So function indicates IAE the IAE bit is set in the SM field. And finally the BIP-8 even parity code is calculated and inserted in the SM field.

4.11.3   The Connection Function: OTUk_C

There is no connection function present in this layer.

4.11.4   The Sink Termination Function: OTUk_TT_Sk

This function extracts the OTUk section specific OAM from the SM field of the OTU-OH. It will use this information for checking the connectivity in the OTUk layer, the OTUk section status, and performance monitoring. The detailed description of the SM field is provided in Table 4-13.

If the API in the TTI bytes of the SM field is not equal to the ExTI, this will be reported to the EMF as TIM defect. A TSF signal is output to the succeeding OTUk/ODUk_A_Sk function and BDI is sent to the co-located ODUk_TT_So function for transfer to the far-end. The received BDI is reported to the EMF for fault localization.

The received BIP-8 code is compared to the calculated BIP-8. The number of detected block-errors is used for near-end performance monitoring, and sent to the co-located ODUk_TT_So function for transfer to the far end in the BEI bits. The block-error count received via the BEI bits is used for local far-end performance monitoring. If a *Degraded signal* (DEG) defect is detected a TSD signal is output to the succeeding OTUk/ODUk_A_Sk function.

4.11.5   The Sink Adaptation Function: OTUk/ODUk_A_Sk

This function extracts the ODUk signal from the OTUk signal. The ODUk clock is generated by dividing the received OTUk clock by a factor 255/239. If the function is provisioned "locked" the output ODUk_CI signal is replaced by a LCK signal. If a TSF signal is received from the preceding OTUk_TT_Sk function the function will output an all-ones AIS signal as well as an SSF signal. If a TSD signal is received from the preceding OTUk_TT_Sk function the function will output an SSD signal.

4.11.6   The OTUk Section Overhead OTU-OH

Table 4.13 provides a description of the OTUk (k = 1, 2, 3) section specific overhead fields. The overhead is located in [1, 8 ... 14] of the OTUk frame structure.

TABLE 4-13. OTUk section overhead fields OTU-OH

Field	Description
FAS	*Frame Alignment Signal* (FAS) field—consists of six bytes: OA1 Three bytes containing the value "1111 0110" located in [1, 1 ... 3] OA2 Three bytes containing the value "0010 1000" located in [1, 4 ... 6]
MFAS	*Multi-Frame Alignment Signal* (MFAS) field—a single byte located in [1, 7] containing a counter, each OCh frame the counter is incremented, counting: 0, 1, ... , 254, 255, 0, 1, ...

SM	*Section Monitoring* (SM) field—consists of three bytes:
	
	TTI *Trail Trace Identifier.* A 64-byte string containing:
	      —a 16-byte source *Access Point Identifier* (API),
	      —a 16-byte destination API and
	      —a 32-byte operator specific field.
	
	BIP-8 *8-Bit Interleaved Parity* code calculated over the OPUk frame i (column 15—3624) and inserted in the BIP-8 byte in frame i + 2.
	
	status bits [1 ... 4]—*Backward Error Indication* (BEI) block count and
	                —*Backward Incoming Alignment Error* (BIAE) indicator
	      bit [5] —*Backward Defect Indication* (BDI) if set to "1"
	      bit [6] —*Incoming Alignment Error* (IAE) is set to "1"
	      bits [7, 8] —*Reserved* (RES) for future use

GCC0	*Generic Communications Channel* (GCC) field—two bytes provide a clear channel between any two network elements that have access to the OTU-OH.
RES	*Reserved* (RES) field—two bytes, these are for future use.

4.12   THE OTN OPTICAL CHANNEL LAYER (OCh)

The OCh layer generally has the OTUk layer as its client layer. It either has the OMS layer as its server layer or, in case of reduced capability, the OPS layer as its server layer. Figure 4-32 shows how the layers are connected and how the signals are processed by each atomic function.

The OCh layer can also have a CBRx layer or an STM RSn layer as its client layer.

The OCh layer has full functionality and uses non-associated overhead, supported by the *OTM Overhead Signal* (OOS), while the OCHr layer has reduced functionality and uses no overhead. As Figure 4-32 shows, the OCh and OCHr layers have different trail termination functions and a common adaptation function. The connection function is only specified for the OCh layer.

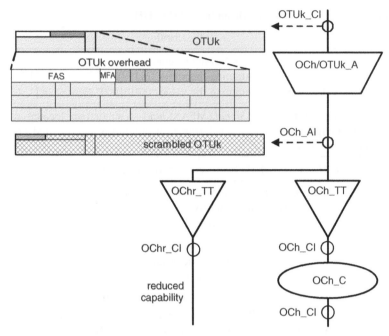

Figure 4-32. OCh layer

4.12.1   The Source Adaptation Function: OCh/OTUk_A_So

This function inserts the FAS and MFAS values in the OTU-OH as described in Table 4-13. To extend the range of the optical signal by correcting bit-errors or to monitor the bit-error rate a special adaptation function is used that inserts a *Forward Error Correction* (FEC) code: the OCh/OTUk-a_A_So function. The OTUk uses a 16-byte interleaved Reed-Solomon code: blocks of 239 bytes are protected by 16 byte FEC codes. The adaptation function not supporting FEC is identified as: OCh/OTUk-b_A_So. The OTUk signal is scrambled by a frame synchronous scrambler with a generating polynomial of $1 + x + x^3 + x^{12} + x^{16}$. The scrambler starts at the first bit after the FAS field in the OTU-OH and stops at the last bit of the OTUk frame structure. The scrambler is used to prevent long sequences of "1"s or "0"s in the output bit-stream.

4.12.2   The Source Termination Function: OChr_TT_So (reduced)

This function conditions the OCh signal for transmission over the optical medium. The characteristic information OChr_CI has reduced capability because it does not support OCh overhead. The signal is sent to the OPS layer.

4.12.3   The Source Termination Function: OCh_TT_So

This function conditions the OCh signal for transmission over the optical medium. The characteristic information OCh_CI supports both OCh payload and OCh overhead. The signal is sent to the OCh_C function.

4.12.4   The Connection Function: OCh-C

This function provides the connectivity in the OCh layer. In case an output is not connected to an input an *Open Connection Indication* (OCI) signal will be output in the OOS.

The OCh_C function can provide 1 + 1 *Sub-Network Connection* (SNC) protection. The protection mechanisms are described in Chapter 7. Protection switching can be initiated by SSF and SSD signals received from the server layer.

An SSF signal present at the input is transferred to the same output as the payload data.

4.12.5   The Sink Termination Function: OCh_TT_Sk

This function recovers the OCh payload signal and OCh overhead from the optical signal. It will detect and report a LOS defect. The function extracts also the OCh overhead from the OOS. In the overhead it will detect and report the OCI defect and the *Forward Defect Indication* for *payload* (FDI-P) and *overhead* (FDI-O) defects. LOS, OCI, or FDI-P are reported to the succeeding OCh/OTUk_A_Sk function as TSF-P and FDI-O as TSF-O.

4.12.6   The Sink Termination Function: OChr_TT_Sk (reduced)

This function recovers the OCh payload signal from the optical signal. It will detect and report the LOS defect. The detected LOS is transferred to the succeeding OCh/OTUk_A_Sk function as the TSF-P signal.

4.12.7   The Sink Adaptation Function: OCh/OTUk_A_Sk

This function processes the OCh signal to recover the OTUk frame structure. First the clock signal is recovered. Then the alignment process will generate the frame start signal or detect the LOF defect. LOF is reported to the EMF. Next the recovered data signal is descrambled. In the OCh/OTUk-a_A_Sk function the FEC decoder checks for possible bit-errors. The FEC code can either correct up to 8 bit-errors or detect up to 16 bit-errors. The OCh/OTUk-b_A_Sk function does not support FEC coding. Both functions will then recover the multiframe start or detect a LOM defect. The LOM will be reported to the EMF. LOF or LOM defects or received TSF-P will be reported to the succeeding OTUk_TT_Sk function as SSF.

4.12.8 The Non-OTN Adaptation Function: OCh/<client>_A

Two non-OTN adaptation functions are currently (2008) defined in [ITU-T Rec. G.798], the CBRx to OCh adaptation function OCh/CBRx_A and the SDH RSn to OCh adaptation function OCh/RSn_A.

4.12.8.1 The CBRx to OCh Adaptation Function: OCh/CBRx_A The OCh/CBRx_A_So function maps the CBRx client signal (x = 2G5, 10G, or 40G) into the OCh signal. The OCh/CBRx_A_Sk recovers the CBRx signal from the OCh signal. The supported bitrates comply with the SDH STM-N (N = 16, 64, and 256). If a TSF signal is received from the preceding OCH_TT_Sk function it outputs a generic AIS signal and as SSF towards the succeeding CBRx layer function.

4.12.8.2 The SDH RSn to OCh Adaptation: OCh/RSn_A The OCh/RSn_A_So function maps the RSn signal into the OCh signal. The OCh/RSn_A_Sk funtion recovers the RSn signal from the OCh signal. Note that n = 16, 64, or 256. The alignment process will generate the frame start signal or detect a LOF defect. If the function detects a generic AIS signal, a LOF defect, or receives a TSF signal from the preceding OCH_TT_Sk function it outputs an AIS signal and an SSF towards the succeeding RSn_TT_Sk function.

4.13 THE OTN OPTICAL MULTIPLEX SECTION LAYER (OMSn)

The OMS layer has the OCh layer as its client layer. It has the OTS layer as its server layer. Figure 4-33 shows how the layers are connected.

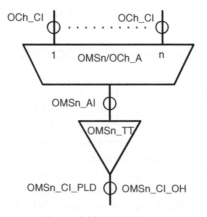

Figure 4-33. OMS layer

In the OMSn layer n OCh traffic wavelengths are multiplexed in the OMSn payload, i.e., the OMSn_CI_PLD. The OMS and OCh overhead information OMSn_CI_OH is transferred in the OOS. The OOS can also contain general management communications. The OMSn trail can be protected by creating a trail protection sublayer according to the functional modeling rules. This is described in Section 7.3.1.1 of Chapter 7.

4.13.1 The Source Adaptation Function: OMSn/OCh_A_So

This function will, if required, precondition the optical OCh payload signal before multiplexing it with other optical OCh signals. Up to n OCh signals can be multiplexed. The optical multiplex signal itself may also need preconditioning to match it to a specific OTM-n interface. This function will also multiplex the individual OCh overhead into the OOS.

4.13.2 The Source Termination Function: OMSn_TT_So

This function monitors the received OMSn adapted signal to detect a missing payload signal. This defect is reported by setting the *Payload Missing Indication* (PMI) in the OOS. The function will also insert in the OMS-OH of the OOS the BDI-P and BDI-O indicators received from the co-located OMSn_TT_Sk function.

4.13.3 The Connection Function

There is no connection function present in this layer.

4.13.4 The Sink Termination Function: OMSn_TT_Sk

This function monitors the received OMSn PLD signal to detect a *Payload Loss of Signal* (LOS-P) defect. The function also extracts and processes the FDI-P, FDI-O, BDI-P, BDI-O, and PMI information from the OMS-OH of the OOS. The LOS-P or FDI-P defects or received SSF-P are reported to the succeeding OMSn/OCh_A_Sk function as TSF-P and as BDI-P to the co-located OMSn_TT_So function for transfer to the far-end.

An FDI-O defect or a received SSF-O is reported to the succeeding OMSn/OCh_A_Sk function as TSF-O and as BDI-O to the co-located OMSn_TT_So function for transfer to the far-end.

The TSF-P and TSF-O will be used for near-end performance monitoring and the BDI-P and BDI-O received from the remote OMSn_TT_So function are used for far-end performance monitoring.

4.13.5 The Sink Adaptation Function: OMSn/OCh_A_Sk

This function will, if necessary, postcondition the optical multiplex signal before demultiplexing it into the n individual optical OCh payload signals. The

OOS is also demultiplexed and distributed to the individual OCh OOS. The TSF-P and TSF-O signals received from the preceding OMSn_TT_Sk function will be forwarded as SSF-P and SSF-O to the OCh_C function and also inserted as FDI-P and FDI-O into each of the OCh-OH of the OOS. Each OCh signal may also need post-conditioning before it is output as OCh_CI.

4.14 THE OTN OPTICAL TRANSMISSION SECTION LAYER (OTSn)

The OTSn layer has the OMSn layer as its client layer. It has the optical fiber as its server layer. Figure 4-34 shows how the layers are connected. The OTSn layer provides the optical signal that can be transported over the fibers in the OTN. It can verify the connectivity of the fibers.

The OTS, OMS, and OCh overhead information is transferred in an *Optical Supervisory Channel* (OSC), i.e., a wavelength dedicated to carry the overhead. The OSC can also transfer general management communications.

4.14.1 The Source Adaptation Function: OTSn/OMSn_A_So

This function will, if necessary, precondition the optical OMSn signal to match it to a specific OTM-n interface. The OOS is passed transparently.

4.14.2 The Source Termination Function: OTSn_TT_So

This function monitors the adapted OTSn signal to detect and report the PMI defect. It will be inserted in the OTS-OH of the OOS together with the BDI-P and BDI-O indicators received from the co-located OTSn_TT_Sk function and the TTI with the API provisioned by the EMF. The logical OOS is mapped into the OSC information structure.

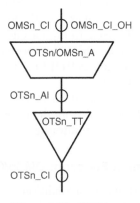

Figure 4-34. OTS layer

It may be required to provide *Automatic optical Power Reduction* (APR) for eye safety. APR can be triggered by the APR signal received from the co-located OTSn_TT_Sk function or it may be provisioned by the EMF.

4.14.3 The Connection Function

There is no connection function present in this layer.

4.14.4 The Sink Termination Function: OTSn_TT_Sk

This function extracts the OTS payload signal and the OSC signal from the received OTSn signal. It will detect LOS-P and LOS-O defects. The function extracts and processes the BDI-P, BDI-O, PMI, and TTI information from the OTS-OH of the OOS in the OSC structure. The API in the extracted OTS-TTI field is compared to the expected API, if there is a difference this will be reported to the EMF as a TIM defect.

LOS-P or TIM defects are reported to the succeeding OTSn/OMSn_A_Sk function as TSF-P and as BDI-P to the co-located OTSn_TT_So function for transfer to the far-end. LOS-O or TIM defects are reported to the succeeding OTSn/OMSn_A_Sk function as TSF-O and as BDI-O to the co-located OTSn_TT_So function for transfer to the far-end. The LOS-P, LOS-O and TIM defects are used for near-end performance monitoring and the BDI-P and BDI-O received from the remote OTSn_TT_So function are used for far-end performance monitoring.

If APR is supported the OTSn_CI signal is monitored and if the triggering criteria are satisfied the APR signal is sent to the co-located OTSn_TT_So function for transfer to the far-end.

4.14.5 The Sink Adaptation Function: OTSn/OMSn_A_Sk

This function will, if necessary, postcondition the optical OTSn signal. The TSF-P and TSF-O signals received from the preceding OTSn_TT_Sk function are inserted as FDI-P and FDI-O into the OMS-OH of the OOS and will be forwarded as SSF-P and SSF-O to the client OMSn layer. If TSF-P is active the output payload signal OMSn PLD is switched off.

4.15 THE OTN OPTICAL PHYSICAL SECTION LAYER (OPS)

The OPSn (n = 0, 16) layer has the OCh with reduced capability (OChr) as its client layer and it has the optical fiber as its server layer as shown in Figure 4-35. The OPSn layer provides the optical signal that can be transported over the fibers in the OTN. It can verify the connectivity of the fibers. Currently only the transport of a single OChr or 16 OChr is supported, as indicated in Figure 4-35.

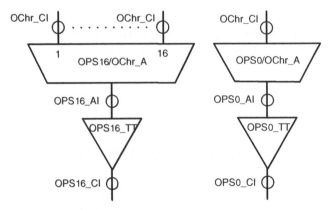

Figure 4-35. OPS layer

The transfer of nonassociated overhead information is not supported.

The physical characteristics of OTM-n interfaces are defined in [ITU-T Rec. G.959.1].

4.15.1   The Source Adaptation Function: OPSn/OChr_A_So

This function will, if necessary, precondition the optical OChr signals to comply with the OTM-n specification.

4.15.2   The Source Termination Function: OPSn_TT_So

This function forwards the adapted OPSn signal without processing to the fiber connected to its output.

4.15.3   The Connection Function

There is no connection function present in this layer.

4.15.4   The Sink Termination Function: OPSn_TT_Sk

This function will detect the LOS-P defect and report it to the succeeding OPSn/OChr_A_Sk function by sending a TSF-P signal. LOS-P will also be reported to the EMF and is used for near-end performance monitoring.

4.15.5   The Sink Adaptation Function: OPSn/OChr_A_Sk

This function will postcondition the optical OChr signals before they are output. The TSF-P signal received from the preceding OPSn_TT_Sk function is forwarded as SSF-P to each of the client OCh layer outputs.

CHAPTER 5

PACKET TRANSPORT NETWORK MODELING: EXTENDING THE MODEL TO NGN

The functional model of the *Packet Transport Network* (PTN) is similar to that of the SDH and OTN networks. The same layering and partitioning principles can be applied. It has been designed to continue to use the functional modeling in the *Next Generation Network* (NGN). The fundamental PTN model is shown in Figure 5-1.

Three distinctive layers can be identified in the PTN:

- The *Packet Transport Channel* (PTC) layer. In this layer the client signals are adapted to the characteristic packet format of the PTN. The characteristic information of the client layer can be packet technology signals or TDM signals that are mapped into PTN packets by the adaptation function.
- The *Packet Transport Path* (PTP) layer. This is the transport layer in the PTN that provides the connectivity. Paths can be bundled and transported as a new entity (tunnel) through the network, this creates a new path layer in the PTN model.
- The Packet *Transport Section* (PTS) layer. The section layer provides the link connections that connect the subnetwork connections in the PTP layer. The characteristic information of the server layer can be packet based or TDM based.

The ComSoc Guide to Next Generation Optical Transport: SDH/SONET/OTN,
by Huub van Helvoort
Copyright © 2009 Institute of Electrical and Electronics Engineers

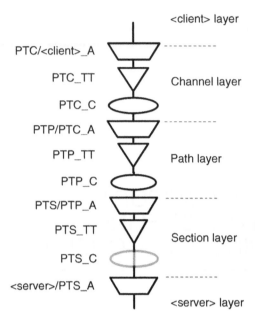

Figure 5-1. PTN functional model

Currently the ITU-T has developed two PTN technologies. It has specified "Transport Ethernet" in the G.80xx series and "Transport MPLS" in the G.81xx series of recommendations.

5.1 THE GENERIC PTN LAYER

Figure 5-2 shows the generic PTN network layer model. The figure shows the adaptation of the <client> layer (client signal) into the PTN layer, the PTN trail termination function where the PTN trail starts and stops, and the PTN connection function taking care of the network connectivity and, if provisioned, the connection protection. For each of the three layers shown in figure 5-1 this generic model can be re-used. In the PTN packets there is no overhead available for the layer-specific OAM. For this purpose special OAM traffic units are inserted into and extracted from the transported traffic stream.

Because the OAM traffic units are present in each layer the layer to which they belong has to be identified. For this purpose *Maintenance Entities* (ME) are specified. A network connection in the PTN runs between two *Maintenance End Points* (MEP). The network connection can be monitored within an ME at *Maintenance Intermediate Points* (MIP).

The adaptation and termination functions provide pro-active maintenance, e.g., for checking the continuity and connectivity of the trail.

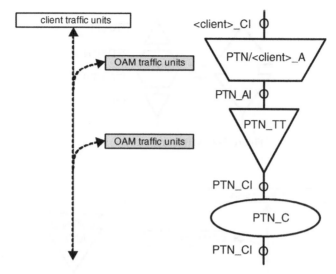

Figure 5-2. PTN network layer model

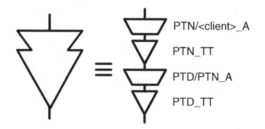

Figure 5-3. MEP compound function

For on-demand maintenance a special function is defined: the *Packet Transport Diagnostic* (PTD) function. The PTD function is similar to a termination function with additional performance monitoring capabilities.

To simplify the functional model compound functions can be used that represent a group of atomic functions. Figure 5-3 shows the compound function of a MEP and which atomic functions it represents. In general the MEP function is used to monitor the network connection.

Figure 5-4 shows the compound function of a MIP and which atomic functions it represents. In general the MIP function is used to monitor the link connection in a particular trail.

Figure 5-5 shows the PTN layer model again with the MEP function replacing the adaptation and termination functions and the location of the MIP function.

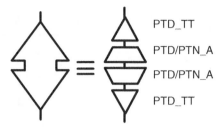

Figure 5-4. MIP compound function

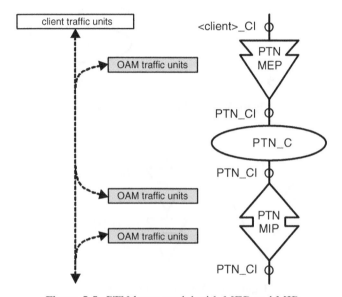

Figure 5-5. PTN layer model with MEP and MIP

5.1.1   The Source Adaptation Function: PTN/<client>_A_So

This function either passes client traffic units transparently (e.g., Ethernet packets) or adds a layer-specific header (e.g., MPLS). The function can multiplex several client traffic streams. At the section layer PTS this function will also insert OAM traffic units that it receives from the management layer, i.e., MCC, and from the timing layer, i.e., SSM. If the <client>_CI contains APS information this information is transferred to the sink side by APS OAM traffic units that are inserted in the traffic stream. When the function is provisioned to block traffic, only LCK OAM traffic units are sent to the sink side. The Maintenance Entity Level (MEL) is provisioned by the EMF and is used to identify the layer that this adaptation function belongs to. The OAM traffic units are described in Table 5-1.

TABLE 5-1. PTN OAM traffic unit description

Byte	Description
CCC	*Continuity and Connectivity Check*—pro-active OAM—used to verify whether the correct network connection has been made between any pair of MEPs. It also provides dual-ended loss measurement at the near-end and the far-end.
LCK	*Lock*—maintenance—used to indicate that traffic is blocked by the operator (and not due to a detected defect).
LMM LMR	*Loss Measurement* (LM), Message and Response—on-demand—used for single-ended LM. Traffic units are counted at ingress and at egress and the counter values are compared to determine severely errored seconds
LBM LBR	*LoopBack* (LB), Message and Response—on-demand—used to verify the connectivity between a MEP and a MIP, or its peer MEP.
DMM DMR	*Delay Measurement* (DM), Message and Response—on-demand—used for two-way measurement of the frame delay and frame delay variation between a pair of MEPs.
1DM	*One-way Delay Measurement*—on-demand—used for one-way DM.
LTM LTR	*Link Trace* (LT), Message and Response—on-demand—can be used: a) to retrieve the adjacency relationship between a MEP and a peer MEP or MIP, i.e., the sequence of MIPs between the source MEP and the target MIP or MEP. b) for fault localization by comparing the retrieved sequence of MIPs with the expected one. The difference in the sequences provides information about the fault location.
APS	*Automatic Protection Switching*—maintenance—used by the APS protocol improve the availability (see also Chapter 7).
TST	*Test*—on-demand—used to perform one-way in-service or out-of-service diagnostics tests, e.g., verifying bandwidth throughput, frame loss, bit errors, etc. The TST OAM traffic units are inserted with a specified throughput, frame size, and transmission pattern.
FDI AIS	*Forward Defect Indication* and *Alarm Indication Signal*—maintenance—used for alarm suppression in the client layers when a defect has been detected in a server layer.
BDI	*Backward Defect Indication*—maintenance—used to inform the peer MEP that a service affecting defect has been detected; this indication can also be transferred by an RDI flag in the CCC OAM traffic unit.
MCC	*Management Communications Channel*—maintenance—used to transfer management information between network elements for remote access.
SSM	*Synchronization Status Message*—maintenance—used to transfer network timing information, e.g., timing quality, between nodes in a network.
EXP	*EXPerimental*—maintenance—used temporarily. Its purpose is undefined. It is expected to be restricted to a single operator's domain.
VSP	*Vendor SPecific*—maintenance—used by a vendor. Its purpose is undefined. It is expected to be restricted to a single vendor equipment.

5.1.2 The Source Termination Function: PTN_TT_So

This function passes client traffic units transparently. It will insert the PTN trail-specific CCC OAM traffic units. The Maintenance Entity Level (MEL) is provisioned by the EMF and is used to identify the layer that this termination function belongs to. A *Remote* or *Backward Defect Indication* (RDI/BDI) received from the co-located PTN_TT_Sk function is either flagged in the CCC OAM or inserted as a BDI OAM traffic unit. The OAM traffic units are described in Table 5-1.

5.1.3 The Source MEP Function: PTN_MEP_So

This function has the same functionality as the PTN source adaptation and termination functions as is shown in Figure 5-4. In addition it contains a PTN diagnostic function (PTD) that provides additional monitoring functionality. The PTD/PTN_A function is empty; it is only present to comply with the modeling rules. The MEP function inserts: LMM and LMR, LBM and LBR, DMM and DMR, 1DM, LTM and LTR, and TST OAM traffic units with the provisioned MEL. The OAM traffic units are described in Table 5-1.

5.1.4 The Connection Function: PTN_C

This function provides the connectivity within a PTN layer network. If provisioned, this function will also provide the trail protection of the PTN trails as described in detail in Chapter 7. Protection switching can be initiated by the *Server Signal Fail* (SSF) condition that is received from the server layer; the *Automatic Protection Switching* (APS) protocol is used to align both sides of the protecting entities. An SSF condition present at the input is transferred to the same output as the traffic units.

In the section layer PTS the connection function is generally a single connection.

5.1.5 The MIP Function: PTN_MIP

The Maintenance Entity Level (MEL) is provisioned by the EMF and is used to identify the layer that this MIP function belongs to. The MIP sink function will extract LTM and LBM OAM traffic units with the provisioned MEL. The co-located MIP source function will respond by inserting in the reverse direction the appropriate LTR and LBR OAM traffic units. The OAM traffic units are described in Table 5-1.

5.1.6 The Sink MEP Function: PTN_MEP_So

This function has the same functionality as the PTN sink adaptation and termination functions as is shown in Figure 5-4. In addition it contains a PTN diagnostic function (PTD) providing additional monitoring functionality. The MEP function extracts: LMM and LMR, LBM and LBR, DMM and DMR, 1DM, LTM and LTR, and TST OAM traffic units with the provisioned MEL. OAM traffic units with a different MEL are passed through. The OAM traffic units are described in Table 5-1.

5.1.7   The Sink Termination Function: PTN_TT_Sk

This function terminates the PTN trail. The Maintenance Entity Level (MEL) is provisioned by the EMF. It will extract the PTN trail-specific CCC as well as the LCK and possible FDI OAM traffic units if they have the correct MEL. The CCC OAM is used to detect a *Loss of Continuity* LOC defect. CCC OAM traffic units that have a lower level MEL will also be extracted; it is used to determine an *unexpected MEL* (UML), a *mismerge* (MMG), or an *unexpected MEP* (UMP) defect. All other traffic units are passed through unless the traffic is blocked due to the detected UML, MMG, or UMP defects. The OAM traffic units are described in Table 5-1.

If an SSF signal is received from the server layer, this condition is transferred to the following adaptation function as TSF and a RDI/BDI signal is sent to the co-located PTN_TT_So function for transfer to the far-end. TSF is also generated when a LOC, UML, MMG, or UMP defect is detected.

A received RDI/BDI is sent to the EMF to be used for fault handling.

5.1.8   The Sink Adaptation Function: PTN/<client>_A_Sk

This function either passes client traffic units transparently (e.g., Ethernet packets) or removes the layer-specific header (e.g., MPLS). The Maintenance Entity Level (MEL) is provisioned by the EMF. APS OAM traffic units are extracted from the traffic stream and the APS information transferred to the connection function that follows. At the section layer PTS MCC OAM traffic units are extracted and sent to the management layer, and SSM OAM traffic units are extracted and sent to the timing layer. When the function is provisioned to block traffic, only LCK OAM traffic units are generated. If a TSF condition is received from the preceding termination function FDI/AIS OAM traffic units are inserted and forwarded by the SSF signal to the next function. The OAM traffic units are described in Table 5-1.

5.2   THE PTN-SPECIFIC OAM TRAFFIC UNITS

Table 5.1 provides an overview and description of the PTN OAM traffic units.

CHAPTER 6

FREQUENCY JUSTIFICATION: POINTERS AND STUFFING

Because client signals may have a bit-rate that is not synchronous with the clock of the server container that will transport the client signal, it is necessary to provide a mechanism that preserves the client clock during the transport. This ensures that the bit-rate of the client signal at the egress is exactly the same as the bit-rate at the ingress, and no data is lost during transport.

There are three mechanisms used for justification of the difference in frequency:

- **Pointer processing.** This mechanism is used by the multiplexing processes in SDH. One or more client virtual containers (VC-m) are *"floating"* in the payload area of an administrative/tributary unit (AUn/TUn, n ≥ m), the pointer *"points"* to the start of the client structure in the payload area. By changing the pointer value differences in frequency can be accommodated.

- **Bit stuffing.** This mechanism is used by the mapping processes in SDH. When client signals are mapped into the payload area of SDH containers some space is available in the payload area for frequency justification. The justification process uses control bits and a justification opportunity bit/byte, i.e., the "stuff" bit/byte, to compensate the frequency difference.

The ComSoc Guide to Next Generation Optical Transport: SDH/SONET/OTN,
by Huub van Helvoort
Copyright © 2009 Institute of Electrical and Electronics Engineers

- **Positive and negative justification.** This mechanism is used by both the mapping and the multiplexing processes in OTN. The justification process uses the *Justification OverHead* (JOH), which is part of the OPUk overhead, to control the justification and to provide justification opportunities to compensate the frequency difference between client and server signals.

These three mechanisms are described in more details in the following sections.

6.1 POINTER PROCESSING

In SDH the use of pointers provides a methodology to allow a flexible and dynamic alignment of the VC-n within an AU-n or TU-n frame. The pointer mechanism was designed specifically for the application in SDH.

Dynamic alignment allows the VC frame to "float" within the AU/TU frame. This means that the pointer is able to accommodate differences, in both frame phase and frame rate, between the VC frame and AU/TU frame. By its design the pointer mechanism can accommodate a clock accuracy of ±320 ppm (parts per million). Note that the SDH STM-N {STS-N} network clock has an accuracy of ±4.6 ppm.

Figure 6-1 shows a VC-4 frame (indicated by the hatching) that is "floating" in the payload area of successive AU-4 frames (indicated by the gray color). The AU-4 frame is part of an STM-1 structure and consists of the AU-4 pointer located in row 4, columns 1 ... 9, and the AU4 payload area located in columns 10 ... 270.

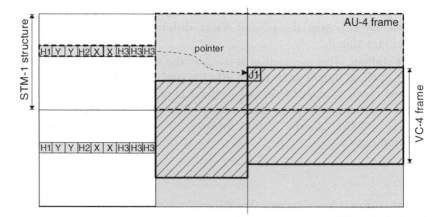

Figure 6-1. A VC-4 "floating" in AU-4 frames

The AU-4 pointer value is a binary number with a range of 0 to 782 which indicates the offset, in three-byte increments, between the pointer and the first byte of the VC-4. A pointer value of 0 indicates that the J1 byte of the VC-4 is located in the byte that immediately follows the last of the three H3 bytes. The value of the Y bytes is "1001ss11" (the bits "ss" are unspecified, and ignored at reception) and the value of the X bytes is "11111111" to indicate that the VC4 is in fact a concatenation of three VC-3s. This is very clear in SONET where an STS-3c SPE, i.e., a VC-4, is the contiguous concatenation of three STS-1 SPEs, i.e., three VC-3s.

Figure 6-2 shows how three VC-3 {STS-1 SPE} are "floating" in the payload area of three related AU-3, the three AU-3 frames are byte/column interleaved multiplexed into an STM-1 {OC3} structure.

The pointer of the first AU-3 is located in row 4, columns 1, 4 and 7. The successive columns of the payload area of the first AU-3 are located in columns 10, 13, 16, ... , 268 of the STM-1 structure. The pointer of the second AU-3 is located in row 4, column 2, 5 and 8, etc.

Figure 6-3 shows a VC-3 frame (indicated by the hatching) that is "floating" in the payload area of successive AU-3 frames (indicated by the gray color). The AU-3 frame is part of an STM-0 {OC-1} structure and consists of the AU-3 pointer located in row 4, columns 1, 2, and 3 and the AU3 payload area located in rows 1 ... 4, columns 4 ... 90 of the STM-0 {OC-1}.

The AU-3 pointer value is a binary number with a range of 0 to 782 which indicates the offset between the pointer and the first byte of its related VC-3. A pointer value of 0 indicates that the J1 byte of its VC-3 is located in the byte that immediately follows the H3 byte in the AU-3 frame.

Similarly Figure 6-4 shows a VC-3 frame that is "floating" in the payload area of the TU-3 frames.

The TU-3 pointer value is a binary number with a range of 0 to 764 which indicates the offset between the pointer and the first byte of the VC-3. A

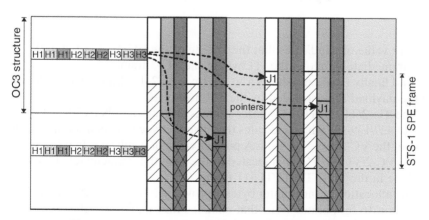

Figure 6-2. Three VC-3 "floating" in three AU-3 frames

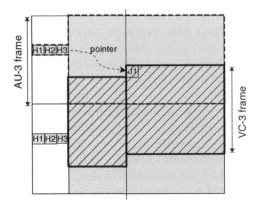

Figure 6-3. A VC-3 "floating" in AU-3 frames

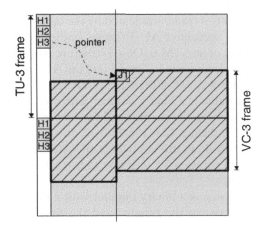

Figure 6-4. A VC-3 "floating" in TU-3 frames

pointer value of 0 indicates that the J1 byte of the VC-3 is located in the byte that immediately follows the H3 byte in the TU-3 frame.

And finally Figure 6-5 shows a VC-m (m = 2, 12, 11) frame that is "floating" in the payload area provided by the TU-m frames.

The TU-2/TU-12/TU-11 pointer value is a binary number within the range of 0 to 427/139/103 which indicates the offset between the pointer and the first byte of the VC-2/VC-12/VC-11. A pointer value of 0 indicates that the V5 byte of the VC-2/VC-12/VC-11 is located in the byte that immediately follows the V2 byte in the TU-2/TU-12/TU-11 frame.

The allocation of the pointer bytes is provided in Table 6-1. PTR1 and PTR2 provide a 16 bit pointer word that contains information for the pointer processing. PTR3 is a byte that is used during frequency justification.

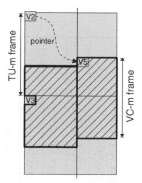

Figure 6-5. A VC-m "floating" in TU-m frames

TABLE 6-1. Pointer byte allocation

	PTR1	PTR2	PTR3
AU-4, AU-3, TU-3	H1	H2	H3
TU-2, TU-12, TU-11	V1	V2	V3

Depending on the pointer action, the PTR3 byte is used to adjust the filling of the input buffers. This byte only carries valid payload information in the event of negative justification, otherwise it is not defined. In case of positive justification the payload byte following the PTR3 byte carries no payload information.

6.1.1 Pointer Word Definition

The administrative unit and tributary unit pointer word provided by the PTR1 and PTR2 bytes is shown in Table 6-2. The pointer word consists of four N bits (bits 1 ... 4), the New Data Flag (NDF), two S bits (bits 5, 6) and the 10 bit pointer value (bits 7 ... 16).

- The NDF is used to indicate that a completely new pointer value and possibly a new size is present and takes effect immediately. If the NDF is enabled is has the value "1001," an NDF enabled is accepted when at least three out of the four bits match the pattern "1001." In this case the pointer value (and size) contains a new value, unless the pointer value is "1111111111" which indicated concatenation. If the NDF is disabled, i.e., during normal operation, it has the value "0110," an NDF disabled is accepted when at least three out of the four bits match the pattern "0110." In this case the pointer value is stable, or is either incremented or decremented as indicated by the "I" and "D" bits. The NDF is invalid when

TABLE 6-2. Pointer word definition

	PTR1					PTR2										Remarks
1	2	3	4	5	6	7	8	9	10	11	12	13	14	15	16	
N	N	N	N	S	S	I	D	I	D	I	D	I	D	I	D	
NDF				Size		10 bit pointer value										
				X	X											AU-4 pointer, max 782 decimal
																AU-3 pointer, max 782 decimal
																TU-3 pointer, max 764 decimal — The S bits are ignored
				0	0											TU-2 pointer, max 427 decimal
				1	0											TU-12 pointer, max 139 decimal
				1	1											TU-11 pointer, max 103 decimal
0	1	1	0			Current pointer value										NDF disabled if ≥3 bits out of 4 match
0	1	1	0			I bits inverted (increment)										Positive justification (majority vote)
0	1	1	0			D bits inverted (decrement)										Negative justification (majority vote)
1	0	0	1			New pointer value										NDF enabled if ≥3 bits out of 4 match
1	0	0	1	X	X	1	1	1	1	1	1	1	1	1	1	Concatenation indication (S bits ignored)
1	1	1	1	1	1	1	1	1	1	1	1	1	1	1	1	AIS indication

one of the values "0000," "0011," "0101", "1010," "1100," or "1111" is detected. If N ($8 \leq N \leq 10$) consecutive invalid NDF are detected a Loss of Pointer defect is detected and reported.

- The two S bits are used to indicate the size of the unit that is using the pointer; the S bits are ignored by higher order pointer processors (AU-4, AU-3 and TU-3).

- The ten-bit pointer value is a binary number that indicates the offset from PTR2 to the first byte of the VC-n/VC-m. The range of the offset is different for each of the tributary unit sizes as illustrated in Table 6-2. The pointer bytes themselves are not counted in the offset calculation. If the pointer processor detects that a majority of the bits in the positions indicated by "I" is inverted AND no majority of the "D" has been inverted AND the S bits have not changed, it will increment the pointer value. A pointer increment can occur only once per four frames. If the pointer processor detects that a majority of the bits in the positions indicated by "D" is inverted AND no majority of the "I" has been inverted AND the S bits have not changed, it will decrement the pointer value. A pointer decrement can occur only once per four frames.

6.1.2 Pointer Justification

Pointer justification is performed either by shifting the "floating" frame by using a single pointer increment or decrement or by providing a completely new pointer value.

In case of a pointer increment or decrement the shift is hitless, i.e., without losing any data. This is achieved by using the H3/V3 bytes that are part of the pointer overhead bytes in case of a decrement, and by skipping the byte immediately following the H3/V3 bytes in case of an increment.

When the original signal is replaced by a signal with a totally different phase, e.g., in case of a protection switch it is the difference in propagation delay of the working path and the protecting path that causes the phase difference, a new pointer value causes a jump in the pointer value.

6.1.3 Pointer Generation

Proper pointer values are generated at the source side if the following set of rules is applied:

a) During stable operation, the pointer "points" to the first byte of the VC-n frame within the AU-m/TU-m frame. That is, the value of the pointer word is the offset from the PTR3 byte in the AU-n/TU-n frame to the J1/V5 byte in the VC-n frame. The value of NDF is set to "0110" (disabled).

b) The value of the pointer word can only be changed by applying rule c), d), or e).

c) When a positive justification is necessary, the value of the current pointer word in the PTR1 and PTR2 bytes is sent with the I-bits inverted and the byte immediately following the PTR3 byte is not used for client data, i.e., it is skipped and filled with fixed stuff. The pointer words in the next frames contain the current pointer value incremented by one. If the current pointer value is at its maximum, the new pointer value is set to zero. No subsequent increment or decrement operation is allowed for at least three frames following this rule. The value of NDF is set to "0110" (disabled).

d) When a negative justification is necessary, the value of the current pointer word in the PTR1 and PTR2 bytes is sent with the D-bits inverted and the PTR3 byte contains client data. The pointer words in the next frames contain the current pointer value decremented by one. If the current pointer value is zero, the new pointer value is set to its maximum. No subsequent increment or decrement operation is allowed for at least three frames following this rule. The value of NDF is set to "0110" (disabled).

e) When a positive or negative justification is necessary that is greater than can be achieved by an increment or decrement, the new pointer value shall be sent and at the same time the NDF is set to "1001" (enabled). The NDF is reset to "0110" (disabled) and sent in the next frames, together with the new pointer value. As soon as the new pointer value is sent the offset it represents is used to write the new client data. No subsequent increment or decrement operation is allowed for at least three frames following this operation.

6.1.4 Pointer Interpretation

Proper interpretation of the pointer values at the sink side is achieved by applying the following set of rules:

a) During stable operation, the pointer "points" to the first byte of the VC-n frame within the AU-m/TU-m frame.

b) A new pointer value different from the current pointer value is ignored unless the same new value is received three times consecutively and NDF is disabled (at least three bits match the pattern "0110"). A new pointer value is also accepted if one of rules c), d), or e) is applicable.

c) A pointer value increment, i.e., a positive justification indicated by inverting the I-bits, is accepted when a majority of the I-bits is inverted and NDF is disabled (at least tree bits match the pattern "0110"). Subsequent pointer values shall be incremented by one.

d) A pointer value decrement, i.e., a negative justification indicated by inverting the D-bits, is accepted when a majority of the D-bits is inverted and NDF is disabled (at least three bits match the pattern "0110"). Subsequent pointer values shall be decremented by one.

e) If the NDF is accepted as enabled, i.e., at least three bits match the pattern "1001," then the pointer value received in the same pointer word is accepted as current pointer value.

Three examples are given to explain the different justifications.

Figure 6-6 shows a negative pointer justification. In this case the H3 byte is used during the decrement of the pointer to carry a byte of the payload. This byte is also referred to as the negative justification opportunity. In Figure 6-6, the three steps are as follows:

Step 1: The H1 and H2 bytes contain a disabled NDF and the current pointer value (e.g., x), the H3 byte is not used.

Step 2: The H1 and H2 bytes contain a disabled NDF and the "D" bits inverted, the H3 byte carries payload information. The content, i.e., the number of bytes, of the VC frame is not affected.

Step 3: The H1 and H2 bytes contain a disabled NDF and the decremented pointer value (e.g., x-1), the H3 byte is not used. This status shall be maintained for at least three frames before a new change is allowed.

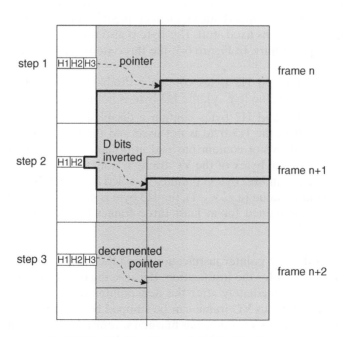

Figure 6-6. Negative pointer justification

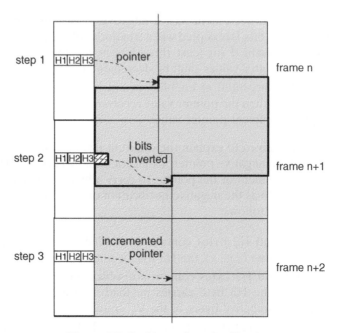

Figure 6-7. Positive pointer justification

Figure 6-7 shows a positive pointer justification. In this case the byte following the H3 byte is skipped during the increment of the pointer. During this frame the byte contains fixed stuff. This byte is also referred to as the positive justification opportunity. In Figure 6-7, the three steps are as follows:

Step 1: The H1 and H2 bytes contain a disabled NDF and the current pointer value (e.g., y), the H3 byte is not used.

Step 2: The H1 and H2 bytes contain a disabled NDF and the "I" bits inverted, the H3 byte is not used and the byte following the H3 byte does not contain payload information. The content, i.e., the number of bytes, of the VC frame is not affected.

Step 3: The H1 and H2 bytes contain a disabled NDF and the incremented pointer value (e.g., y + 1), the H3 byte is not used. This status shall be maintained for at least three frames before a new change is allowed.

Figure 6-8 shows a pointer justification using the NDF. In this case the H3 byte cannot be used for the justification. Because the transmission of the new VC frame starts immediately after the new pointer value is recognized, the transfer of the previous VC frame can be stopped before it is finished or the transfer of the previous VC frame has finished several bytes before the new VC frame transfer is started. This will be experienced as a hit in the client layer.

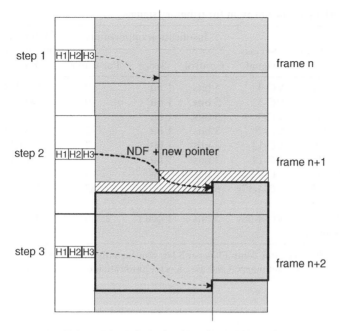

Figure 6-8. Pointer justification with NDF

Step 1: The H1 and H2 bytes contain a disabled NDF and the current pointer value (e.g., x), the H3 byte is not used.

Step 2: The H1 and H2 bytes contain an enabled NDF and a new pointer value (e.g., y), the H3 byte is not used for justification. If y < x the current VC frame is truncated and if y > x there is a gap between the current VC frame and the next.

Step 3: The H1 and H2 bytes contain a disabled NDF and the new pointer value (e.g., y), the H3 byte is not used. This status shall be maintained for at least three frames before a new change is allowed.

6.2 BIT STUFFING

This mechanism is used in SDH for mapping constant bit-rate tributary signals, e.g., PDH and OTN signals. The bit stuffing mechanism was already used in PDH multiplexing where it was designed to accommodate clock accuracies of up to ±50 ppm.

Packet-based tributary signals are mapped without using this justification mechanism; they either adapt their client bit-rate to the available server bit-rate (ATM), or insert idle packets (GFP) to match the server bit-rate. The tributary signals are mapped into the payload area of the virtual containers

TABLE 6-3. Mapping justification for tributary signals

Client signal	Bit-rate in kbit/s	Server signal	Justification opportunity			Minimum bit-rate in kbit/s	Maximum bit-rate in kbit/s
			Control	S size	Per frame		
E4	139,264	VC-4	5 bits	1 bit	9	139,248	139,320
DS3	44,736	VC-3	5 bits	1 bit	9	44,712	44,784
E3	34,368	VC-3	5 bits	1 bit	6	34,344	34,392
DS2	6,312	VC-2	3 bits[1]	1 bit	2	6,304	6,320
E1	2,048	VC-12	3 bits[1]	1 bit	2	2,046	2,050
DS1	1,544	VC-11	3 bits[1]	1 bit	2	1,542	1,546
FDDI	125,000	VC-4	5 bits	1 bit	9	124,960	125,032
ODU1	2,498,775 (approx.)	VC-4-17v	5 bits	1 octet[2]	765	2,496,960	2,545,920
ODU2	10,037,273 (approx.)	VC-4-68v	5 bits	1 octet[2]	2,340	10,033,920	10,183,680

Note 1: these signals have three control bits per S bit, allowing only single C bit errors.
Note 2: these signals are octet mapped and use an octet justification opportunity (S).

and fixed stuffing bits and/or bytes are used to completely fill the payload area. Some of the stuffing bits are used by the justification process.

The process requires control bits (C) to indicate whether a justification is required and justification opportunity bits (S) to provide extra bits for the mapping. Only negative justification is supported, this means that the number of bits allocated for the mapping of the tributary signal in the payload area is sufficient to transport this signal at its lowest possible bit-rate, and together with the additional justification bits sufficient to transport the signal at its highest possible bit-rate.

Normally the use of the justification opportunity bit (S) is indicated by five justification control bits (C). The value CCCCC = "00000" indicates that the S bit contains tributary data, whereas the value CCCCC = "11111" indicates that the S bit is a justification bit. Majority voting is used to take the justification decision at the sink side; this provides immunity for one or two bit errors in the C bits. The value of the S bit, when used as justification bit, is not defined (normally set to "0").

Table 6-3 provides an overview of the existing mappings of tributary signals, the VC-n used for the mapping, and the minimum and maximum supported bit-rate.

6.3 POSITIVE AND NEGATIVE JUSTIFICATION

Instead of using stuffing bits in the payload area the justification mechanism designed specifically for the OTN uses octets from the OPUk overhead as the

Justification OverHead (JOH). The maximum clock tolerance between an OPUk and a client signal that can be accommodated by this mechanism is ±65 ppm. The OPUk clock itself has an accuracy of ±20 ppm which means that the client signal clock can have an accuracy of up to ±45 ppm. This may cause a problem when Ethernet signals (e.g., 10 GbE WAN PHY) that have a tolerance of ±100 ppm have to be transported over the OTN. Nonstandard solutions for this transport are described in [ITU-T Rec. G.Sup.43].

The JOH consists of three octets used for *Justification Control* (JC) and one octet used for *Negative Justification Opportunity* (NJO). The octet(s) for the *Positive Justification Opportunity* (PJO) are located in the payload area because they are used to decrease the number of transported octets. In the JC octet only bits 7 and 8 are used by the justification process to control the justification opportunities, bits 1 ... 6 are reserved for future use. The value of bits 7 and 8 is validated by taking a majority vote (at least two out of three).

6.3.1   OTN Mapping Justification

The process that asynchronously maps CBRx ($x = 2G5, 10G$ and $40G$) signals into the OPUk payload area uses a single PJO octet. This octet is located in row 4 column 17 of the ODUk structure as shown in Figure 6-9. Any frequency difference between the CBRx client clock and the OPUk server clock is compensated by the $-1/0/+1$ justification scheme.

The justification process at the sink adaptation function interprets the JC, NJO, and PJO octets as shown in Table 6-4.

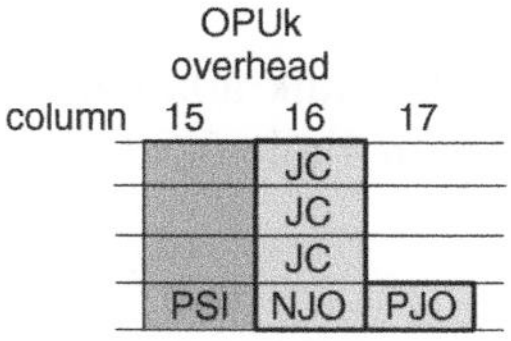

Figure 6-9. Mapping JOH

TABLE 6-4. Mapping JC, NJO, and PJO interpretation

JC	NJO	PJO	Justification interpretation
00	Justification octet	Client octet	None (0)
01	Client octet	Client octet	Negative (−1)
10	Justification octet	Client octet	None (0)—note
11	Justification octet	Justification octet	Positive (+1)

Note: this code will never be generated, it can only be detected due to bit errors

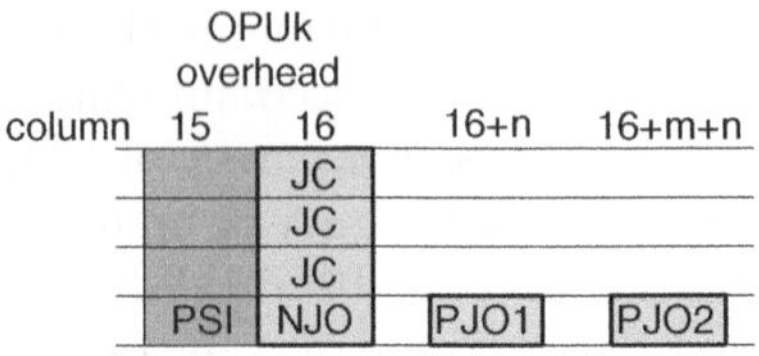

Figure 6-10. Multiplexing JOH

TABLE 6-5. Multiplexing JC, NJO, and PJO interpretation

JC	NJO	PJO1	PJO2	Justification interpretation
00	Justification octet	Client octet	Client octet	None (0)
01	Client octet	Client octet	Client octet	Negative (−1)
10	Justification octet	Justification octet	Justification octet	Double positive (+2)
11	Justification octet	Justification octet	Client octet	Positive (+1)

6.3.2   OTN Multiplexing Justification

The process that asynchronously multiplexes ODUj into an ODUk (k > j) uses two PJO octets. These octets are located in row 4 columns (16 + n) and (16 + m + n) of the ODUk structure as shown in Figure 6-10, where m = 4 or 16 (the number of multiplexed ODUj) and n = 1 ... 4 or 1 ... 16 (identifying the multiplexed ODUj). Any frequency difference between the ODUj client clock and the ODUk server clock is compensated by the −1/0/+1/+2 justification scheme.

The justification process at the sink adaptation function interprets the JC, NJO, and PJO octets as shown in Table 6-5.

CHAPTER 7

PROTECTION MECHANISMS: IMPROVE AVAILABILITY

This chapter describes the protection schemes in a layered network. Protection schemes are used in general to improve the availability of a client signal. The availability is part of the Quality of Service (QoS) agreement between a customer and a service provider. It is also part of the reliability specification of equipment. The different protection schemes are described using functional models.

The schemes are based on the (initial) description of SDH protection in [ITU-T Rec. G.841]. After the introduction of functional modeling, the models are based on the functional description in [ITU-T Rec. G.805] and the equipment specifications in [ITU-T Rec. G.783] and [ITU-T Rec. G.789]. [ITU-T Rec. G.808.1] contains the generic specifications of linear protection and [ITU-T Rec. G.808.2] the generic specification of ring protection.

7.1 AVAILABILITY

Availability is a gauge to measure the ability of a transport network to transfer information under specified performance conditions during a given time interval, assuming that the required resources are provided. This is expressed in the formula:

The ComSoc Guide to Next Generation Optical Transport: SDH/SONET/OTN,
by Huub van Helvoort
Copyright © 2009 Institute of Electrical and Electronics Engineers

$$A\,(\text{vailability}) = \text{Uptime}/(\text{Uptime} + \text{Downtime})$$

The availability of a telecommunications network or a trail (as defined in Section 1.2.1 of Chapter 1) in a transport network is ideally 100%. However, due to human interference (operational and physical) and the physical properties of fibers and electronics, the availability is usually lower if no special measures are taken. A common requirement is 99.999%, also knows as the "*five-nines.*" Availability can also be expressed based on the characteristics of the transport entities, i.e., the *Mean Time Between Failures* (MTBF) and the *Mean Time To Repair* (MTTR):

$$A\,(\text{vailability}) = \text{MTBF}/(\text{MTBF} + \text{MTTR})$$

The availability can be improved either by increasing the MTBF or by decreasing the MTTR. In general, increasing the MTBF is very expensive. The MTTR can be improved by having a replacement (spare part) for the failed or degraded transport entities ready for use. When the replacing is automated the MTTR can be decreased even more.

There are two kinds of replacements. In general:

- **Replacement by protection:** The spare part is available immediately. Preprovisioned additional transport entities between the end points of the protected entity are dedicated to protection. In general this "*stand-by*" or "*protection transport entity*" will replace the "*active*" or "*working transport entity*" to provide transport of the "*normal traffic.*"
- **Replacement by restoration:** The spare part has to be searched for. The spare transport entities are available but need to be provisioned (re-routed) between the end points of the protected entity. When restoration is used in a transport network a certain percentage of the network capacity has to be reserved for the re-routing of the working traffic. In transport networks, i.e., STN, OTN, and PTN the replacement by protection is generally used to guarantee the fife-nines availability.

7.2 PROTECTION ARCHITECTURES

The replacement by protection can be implemented with different levels of complexity. Increased complexity will require less transport entities for protection as shown in the following architectures.

The protection architecture is confined to specific layer in the network. It is possible to nest protection architectures. In this case a hold-off timer should be used to delay the switching actions in the higher layers to allow the lower layers to complete a protection switching action and avoid unnecessary switching.

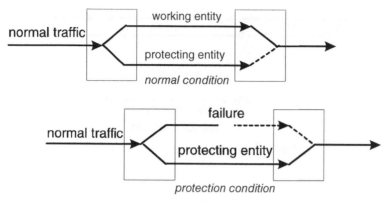

Figure 7-1. $(1 + 1)$ protection

7.2.1 (1 + 1) Protection

This is the simplest form of protection. It is normally referred to as *"one-to-one"* or $(1 + 1)$ protection (Figure 7-1).

In this architecture a single transport entity dedicated to the protection of the normal traffic will replace the working entity in case of a failure. The protection is single ended, i.e., the selection of the working entity or the protecting entity is performed at the sink side only. At the source side the normal traffic is bridged to both the working entity and the protecting entity. The protection is uni-directional, i.e., the protection switch is independent of what happens to the traffic in the opposite direction. For bi-directional $(1 + 1)$ protection switching the *Automatic Protection Switching* (APS) protocol has to be used to synchronize both ends. The protecting entity needs to have the same transport capacity as the working entity: a 100% increase in capacity required to transport the normal traffic. Because the protection switch can be activated immediately after detecting the failure, this switch is very fast.

7.2.2 (1:1) Protection

Under normal conditions, i.e., no protection switching active, the capacity of the protecting entity could be used for transport of traffic that does not need protection and that can even be discarded when protection switching is active: the *"extra traffic."* The architecture that supports this capability is in general referred to as *"one-by-one"* or $(1:1)$ protection and is illustrated in Figures 7-2a and 7-2b.

In this architecture the source end has to know which transport entity the sink end has selected and adjust its bridge accordingly. For this purpose the APS channel has been defined and an APS protocol was developed to align the sink and source end of the transport entity. In general, the $(1:1)$ protection is bi-directional, so both directions will switch to the protecting entity even in

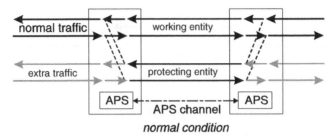

normal condition

Figure 7-2a. (1:1) protection—revertive

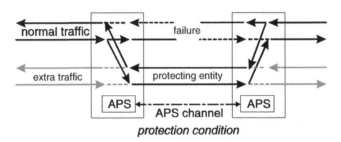

protection condition

Figure 7-2b. (1:1) protection—revertive

case the transport in one direction of the working entity has failed. When the failure of the working entity is repaired it will be detected at the sink side and the bridges and selectors return to their original position. This happens in case of *"revertive"* switching. This switch may cause a traffic hit, i.e., introduce bit-errors in the normal traffic. The traffic hit, even if it is very short, can be prevented by implementing *"non-revertive"* switching; this requires extra bridges and selectors for the extra traffic similar to those of the normal traffic as shown in Figure 7-3.

Note that, in case of *revertive* mode, a wait-to-restore timer should be used to prevent frequent switch operations due to an intermittent defect.

Even though the protecting entity needs to have the same transport capacity as the working entity, the increase in capacity is less than 100% because the protecting entity is used for extra traffic during normal operating conditions. Because the protection switch uses the APS protocol to coordinate source and sink the time to switch will not be as fast as in case of (1 + 1) protection.

7.2.3 (1 : n) Protection

Instead of protecting entities one by one it is also possible to protect n working entities by a single protecting entity. This requires that the failure of a working entity shall be repaired before the protecting entity (statistically) fails. It is

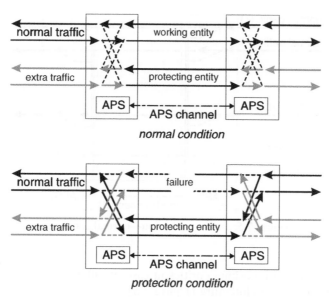

Figure 7-3. (1:1) protection—non-revertive

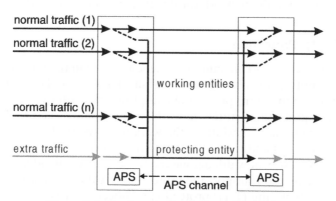

Figure 7-4. (1:n) protection—normal condition

referred to as "one by n" or (1:n) protection and its architecture is shown in Figure 7-4.

The operation of the (1:n) protection is similar to that of the (1:1) protection architecture and is normally used in revertive mode. The capacity required in the network for this type of protection is (100/n)%.

7.2.4 m:n Protection

The (1:n) protection architecture can be extended to have not one but m separate protecting entities. Each of these m entities can be used to replace

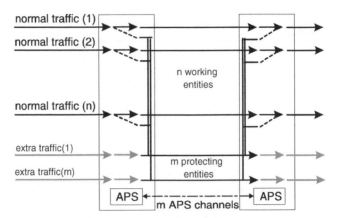

Figure 7-5. (m:n) protection—normal condition

one of the n (m ≤ n) working entities in case of failure. This is the *"m by n"* or (m:n) protection architecture (Figure 7-5).

The operation of the (m:n) protection is similar to that of the (1:1) and (1:n) protection architectures and is normally used in revertive mode. The capacity required in the network for this type of protection is $100 \times (m/n)$ %.

7.2.5 $(1 + 1)^n$ Protection

The $(1:1)^n$ protection architecture is designed for use in the PTN. In this case a transport entity is not a frame structure but a packet flow. The $(1:1)^n$ protection is similar to the (1:n) protection except that the latter requires that the working and protecting entities have the same frame structure. In case of $(1:1)^n$ protection the bandwidth of the working flows can be different and it is required that the bandwidth allocated to the protecting flow is large enough to carry the largest bandwidth among the working flows. In addition the protecting flow shall also transport the protection OAM of all n working flows.

Figure 7-6 shows the $(1:1)^n$ protection architecture, it is normally used in revertive mode. The bandwidth capacity required in the PTN network for this type of protection is $100 \times$ (bandwidth of protecting flow)/(total bandwidth of the working flows) %.

7.3 PROTECTION CLASSES

When SDH and SONET technology were introduced not only the transport capabilities were changed but also new network topologies were introduced based on the experience with PDH networks. PDH has a strong hierarchy in multiplexing and a tree topology. SDH/SONET and OTN networks are more flexible in hierarchy and have a ring topology in many situations. A ring topol-

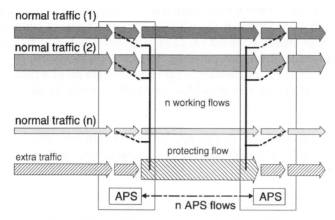

Figure 7-6. $(1:1)^n$ protection—normal condition

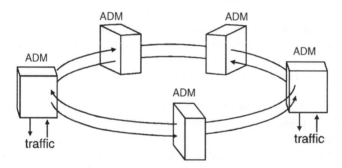

Figure 7-7. Typical ring topology

ogy consists of a number of network elements that allow traffic to be added to the ring, dropped from the ring and transported along the ring. This equipment is generally referred to as *Add/Drop Multiplexers* (ADM). The ADMs are interconnected in a ring configuration as illustrated in Figure 7-7.

The ring topology is by its nature highly suitable for providing protection of the transported traffic and hence improves the availability of network transport entities.

Within the transport network architecture two distinctive types of protection can be identified: *trail protection* and *sub-network protection.*

- Trail protection is used for end-to-end protection in case the transport entity is a trail. The protection switch is based on fault conditions detected by monitoring the trail. Trail protection does not support cascading of trail segments; hence only protection of single fault conditions is supported.

- Sub-network connection protection is used to protect a segment of a path, i.e., a link connection, or the whole path, i.e., a network connection. The protection switch is based on fault conditions detected by monitoring functions added for this purpose. Protected segments can be cascaded, e.g., if each operator chooses to protect the segment of a path in his own domain. Protection is provided for fault conditions in each of the segments.

Both protection classes can be applied in any transport network architecture: a ring topology, a mesh topology, or a mixed topology.

7.3.1   Trail Protection

Trail protection provides end-to-end protection of any path in a layered network. The trail can be protected by adding bridges and selectors at both ends of the trail, and adding an extra trail between these bridges and selectors. The original trail, i.e., the *protected trail,* will have a *working trail,* and a *protecting trail* available to increase the availability. The working trail will use the original sub-network connection and the protecting trail requires its own sub-network connection.

Two types of trail protection will be described:

- Linear trail protection or Path level trail protection. This protection type is used to protect a (linear) network connection between a source and a sink Access Point (AP). A generic specification of linear trail protection is provided in [ITU-T Rec. G.808.1], and technology specific for linear Carrier Class *Ethernet* (ETH) protection in [ITU-T Rec. G.8031], and linear *MPLS_TP Path* (MTP) protection in [ITU-T Rec. G.8131].
- Ring protection or Multiplex section protection or Line level trail protection. This protection type is used to protect all the individual sections in a ring topology. The nodes in the ring can be interconnected by using two optical fibers, one for each direction of transmission, or they can be interconnected by four fibers, one pair of fibers for each direction, to provide more robustness and hence increase the availability. SDH multiplex section protection is specified in [ITU-T Rec. G.841], *Ethernet Physical layer* (ETY) section protection in [ITU-T Rec. G.8032], and *MPLS_TP section* (MTS) protection in [ITU-T Rec. G.8132].

7.3.1.1 Linear Trail Protection   Linear or Path level trail protection can be used on any path in a transport network. It can be applied to any of the following network layers: the SDH Sn layer, the OTN ODUk layer, the transport Ethernet MAC layer (ETH), or the MPLS_TP path layer (MTP). It can also be applied to the SDH MSn section layer and the OTN OMSn layer.

When provisioning the path for the working entity and the protecting entity it is required that they each follow a different physical path through the network from source to sink as depicted in Figure 7-8.

Path level trail protection is modeled by introducing a protection sublayer.

Figure 7-9 shows an unprotected trail in a bi-directional representation. The protection sublayer is inserted as shown in Figure 7-10, for simplicity in a uni-directional way.

The original unprotected trail termination function TT_u, shown in the shaded area in Figure 7-9, is expanded by inserting a protection trail termination function TT_p, a protection connection function C_p and a protection adaptation function A_p according to the functional modeling rules. The C_p function is used to provide the bridge and selector functions required for the protection switch.

Linear VC trail protection switching can operate in a uni-directional or bi-directional manner. Bi-directional protection switching requires an *Automatic Protection Switching* (APS) protocol to synchronize the protections switch at both ends.

The status of the trails in the protection sub-layer, i.e., a trail *Signal Fail* (SF) or *Signal Degrade* (SD), is detected by the unprotected trail termination

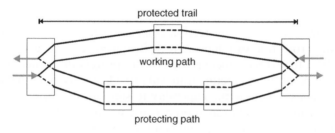

Figure 7-8. Linear trail protection

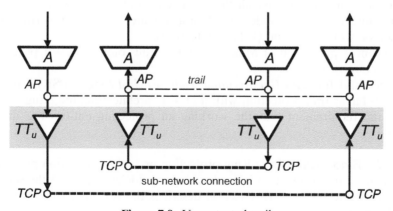

Figure 7-9. Unprotected trail

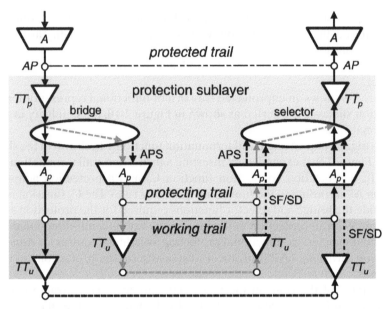

Figure 7-10. Protected trail

TT_u and transferred to the protection connection function C_p to be used as input for the protection switch process. Switching actions can also be initiated by the EMF. If the APS protocol is used by this process the protection adaptation function A_p of the protecting trail will provide access to the APS channel. Figure 7-10 does not show the capability to support an extra traffic trail. In the field this capability is hardly used because it requires that the extra traffic originates and terminates at the same ends as the protected traffic.

Path level trail protection can be applied on individual trails by using the architectures described in Sections 7.2.1 and 7.2.2. It can also be applied to groups of trails; in this case all the architectures in Section 6.2 can be used. Generally the revertive mode is used for all architectures; non-revertive mode is used only in (1 + 1) and (1:1) architectures.

For linear protection of MSn and OMSn trails only the 1 + 1 protection architecture is specified.

In SONET the term *Unidirectional Path Switched Ring* (UPSR) is used to refer to linear trail protection with a (1 + 1) architecture, the "ring" comes from the requirement that the working an protecting entities are diverse routed.

7.3.1.2 Ring Protection Ring protection or Multiplex section protection or line level trail protection is used in ring topologies to protect the sections that constitute the ring. This section describes the two fiber ring protection mechanisms.

In SDH/SONET, ring protection is applied to the higher-order network multiplexes that are used to interconnect a number of nodes in a ring configuration, i.e., an MSn (n = 4, 16, 64, or 256). Because the protecting entities are shared this type of protection switching is also referred to as Shared Protection Ring (SPRing) protection. SONET uses the term *Bi-directional Line Switched Ring* (BLSR) for this protection type. In transport Ethernet and MPLS-TP, ring protection should be applied at the layer that is closest to the physical layer, i.e., an ETY flow or a MTS trail. In OTN ring protection is not yet specified.

The total bandwidth of each section of the ring is shared by the working and protecting entities present in the ring as well as any unprotected entities. The bandwidth of the protecting entities in each section is shared among the working entities of all sections of the ring. The protected trails in the ring are bi-directional and follow the same physical path. Because the protecting bandwidth is shared the total bandwidth in the ring is used more efficiently than in case linear protection would have bee used. Figure 7-11 shows an example of this type of protection in an SDH ring transporting a client signal from node A to node C and vice versa.

This type of trail protection is modeled also by introducing a protection sublayer. The original section to path adaptation function, <sec>/<path>_A, as shown in the shaded area in Figure 7-12, is expanded according to the functional modeling rules.

The result is shown in Figure 7-13. Because the path is bi-directional, both figures are drawn in a bi-directional representation.

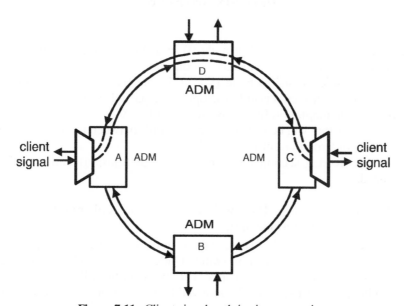

Figure 7-11. Client signal path in ring protection

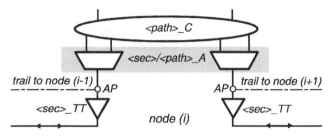

Figure 7-12. Unprotected sections

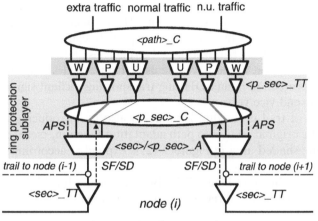

Figure 7-13. Ring protection

Figure 7-12 shows a model of node (i) in a ring topology. Similar to any node in the ring, node i is connected on one side, "*west*," to node (i − 1) and on the other side, "*east*," to node (i + 1). In the path connection function <path>_C signals are either added, dropped, or passed through between two section to path adaptation functions <sec>/<path>_A.

The sections that will be protected are the trails between the section *Access Points* (AP) of two adjacent nodes in the ring.

In the protecting section layer three kinds of transport entities will be recognized:

- Working entities (W). These will transport the normal or protected traffic and ensure a high availability.
- Protecting entities (P). These will be used to transport the protected traffic in case of a protection switch. Under normal conditions these can be used to transport extra traffic. The extra traffic will have the lowest availability because it will be affected by every protection switch in the ring.

- Unprotected entities (U). These will transport traffic that will not be protected and is referred to as *Non-preemptible Unprotected Traffic* (NUT). The availability of the NUT is affected only by a defect in the sections it passes. The reason to identify NUT on the ring is to prevent nesting of protection mechanisms; the NUT may be protected already outside the ring.

The original <sec>/<path>_A function is separated into adaptation functions for each type: designated W, P, and N. For each of these adaptation functions a protection section trail termination function <p_sec>_TT is inserted. The <p_sec>_TT functions are connected to the protection connection function <p_sec>_C. This is then connected via the protection section to section adaptation function <sec>/<p_sec>_A to the existing section termination functions <sec>_TT. See Figure 7-13 for the ring protection model.

The links in the <p_sec>_C function are drawn for the normal, unprotected situation.

Ring protection requires an *Automatic Protection Switch* (APS) protocol to communicate switch requests and status between the control functions of the ring node protection connection functions via a dedicated APS channel.

The status of the protected sections, i.e., a trail *Signal Fail* (SF) or *Signal Degrade* (SD), is detected by the <sec>_TT function and transferred to the <p_sec>_C to be used as input for the protection switch process. Switching actions can also be initiated by the EMF.

The bandwidth capacity of the protecting entities (P_{bw}) of all sections in the ring is determined by the largest bandwidth capacity required by the working entity (W_{bw}) on any of the sections in the ring. The bandwidth capacity of the unprotected entities (U_{bw}) is independent of W_{bw} and P_{bw}. If the maximum bandwidth capacity of every section is assumed to be M_{bw}, then the following rules are applicable:

$$1 \leq W_{bw} \leq (M_{bw}/2), 0 \leq U_{bw} < M_{bw} \text{ and } (2 \times W_{bw} + U_{bw}) < M_{bw}.$$

Figure 7-14 shows how the links in the <p_sec>_C function of node (i) change due to a protection switch after a fault has occurred on the ring section between node (i) and node (i + 1). The working traffic towards node (i + 1) will replace the extra traffic on the protecting entity towards node (i − 1). The traffic received from node (i − 1) that is connected in the <path>_C function to be transmitted to node (i + 1) is now returned to node (i − 1); this is referred to as "*wrapping*."

A similar protection switch will be performed in node (i + 1), or in node (i + 2) if the whole node (i + 1) fails. All the extra traffic in node (i) will be interrupted as well as the NUT to and from node (i + 1) affecting their availability.

Figure 7-15 shows how the links in the <p_sec>_C function of node (i) change due to a protection switch after a fault has occurred on a ring

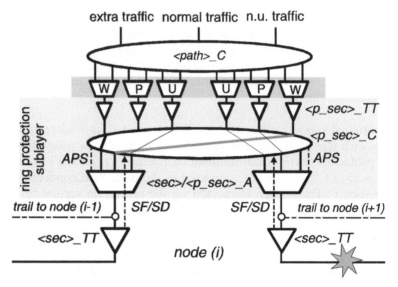

Figure 7-14. Ring protecting failed section

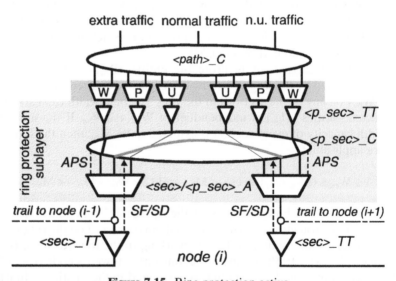

Figure 7-15. Ring protection active

section not adjacent to node (i). This is applicable to all nodes in the ring except the two nodes at both sides of the fault.

When ring protection is active the protecting traffic received from node (i + 1) will be passed through to node (i − 1) and vice versa by connecting the incoming protecting entities to the outgoing protecting entities. All the extra traffic in node (i) will now be interrupted affecting their availability. The NUT is *not* affected in this node.

7.3.1.3 Four Fiber Ring Protection Ring protection or Multiplex section protection or line level trail protection is used in ring topologies to protect the sections that constitute the ring. This section describes the four fiber ring protection mechanisms. It is similar to the two fiber ring protection, but instead of transporting working and protecting entities on a single fiber pair each entity has its own fiber pair.

Figure 7-16 shows an example of ring protection in a four fiber ring transporting a client signal from node A to node C and vice versa. Between two adjacent nodes two fibers are assigned to carry the working entities (solid black) and two fibers are assigned to carry the protecting entities (dashed gray).

This type of trail protection is modelled also by introducing a protection sub-layer. The original section to path adaptation functions, A_W and A_P, as shown in the shaded area in Figure 7-17, is expanded according to the functional modelling rules.

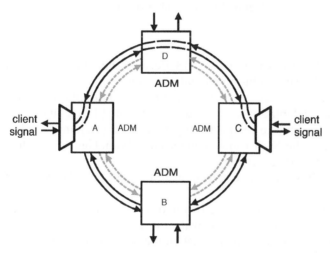

Figure 7-16. Four fiber Ring

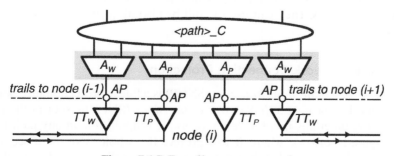

Figure 7-1.7 Four fiber—unprotected

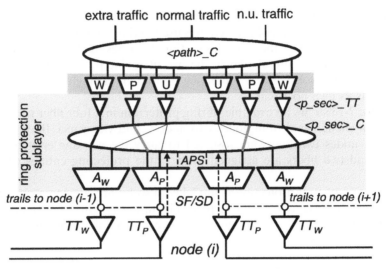

Figure 7-18. Four fiber ring protection

The result is shown in Figure 7-18. Because the path is bi-directional, both figures are drawn in a bi-directional representation.

In the path connection function <path>_C all normal traffic is connected to the working adaptation (A_W) and termination (TT_W), functions, the extra traffic is connected to the protecting adaptation (A_P) and termination (TT_P) functions and the NUT can be connected to both. Figure 7-18 shows a model of node (i) in a ring topology.

Figure 7-19 shows the connections in the <p_sec>_C function after a fault occurs.

When the fibers connecting two nodes fail or a node fails, the two nodes adjacent to the failure will activate the ring protection based on the SF/SD defects as shown in Figure 7-19. In the nodes that are not adjacent to the failure the ring protection will become active based on the APS protocol, the protecting entities are interconnected and the extra traffic is interrupted.

There are no restrictions to the available bandwidth for the working and protecting entities. They can use the full bandwidth provided by the section trails.

7.3.2   Sub-Network Connection Protection

Sub-Network Connection Protection (SNCP) is a linear protection mechanism that is used to protect *Sub-Network Connections* (SNC) present in the path connection functions in a layered network. It is applicable in any path layer in a layered network where it can protect the whole SNC or a segment. However, this is useful only when there are two diverse routes available

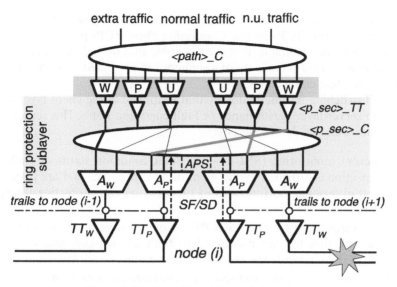

Figure 7-19. Four fiber ring protecting fault

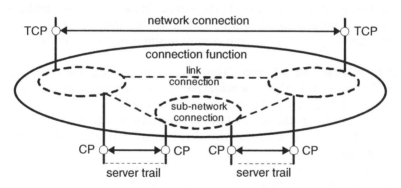

Figure 7-20. Network, sub-network, and link connections

through the network. A generic specification of SNCP is provided in [ITU-T Rec. G.808.1], and technology specific for linear Carrier Class *Ethernet* (ETH) protection in [ITU-T Rec. G.8031] and linear *MPLS_TP Path* (MTP) protection in [ITU-T Rec. G.8131].

The SNC can be protected by adding bridges and selectors at both ends of the trail, and adding an extra SNC between these bridges and selectors. The original SNC, i.e., the *protected SNC*, will thus have a *working SNC* and a *protecting SNC* available to increase the availability. Figure 7-20 shows the definition of a network connection and the subnetwork and link connections in a connection function.

SNCP can be used to protect connections between any pair of *Termination Connection Points* (TCP) and/or *Connection Points* (CP) provided there are two diverse routes available. It is even possible that the sub-network connection being protected is made up of a sequence of protected sub-network connections and link connections.

SNCP switching is a methodology that is applied in the client layer and is based on server layer performance or fault condition status. This server layer performance can be determined by:

- Inherent monitoring (SNC/I)—The fault condition status of each link connection is provided by the monitoring capabilities that are present in the trail termination functions of the working and protecting transport entity.
- Non-intrusive monitoring (SNC/N)—Non-intrusive monitoring termination sink function are connected to both the working and protecting entities, they provide the fault condition status of each link connection.
- Sub-layer trail monitoring (SNC/S)—The (termination) connection points are expanded to insert a sub-layer. The termination functions in this sub-layer will provide the fault condition status of the server layer.
- Test monitoring (SNC/T)—At the source side a test signal is multiplexed with the normal signal and extracted at the sink side. The test signal monitoring provides the fault condition status of the server layer.

7.3.2.1 Inherent Monitoring The inherent monitored SNCP is generally referred to as SNC/I. SNC/I is achieved by provisioning a protecting entity that is routed divers from the original working entity.

It is required that the server layer termination function can provide the server layer performance information that is needed to initiate the protection switching actions.

A functional model of SNC/I is shown in Figure 7-21. The shaded area indicates the added functionality to support SNC/I.

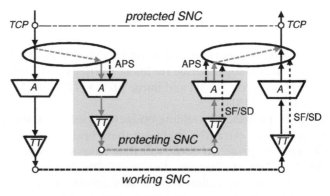

Figure 7-21. SNC/I inherent monitoring

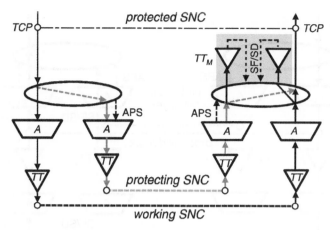

Figure 7-22. SNC/N non-intrusive monitoring

The monitoring capabilities of the existing trail termination TT is used in the SNC/I scheme. The TT monitors the performance of the server layer (sub-)network connection and detects a server signal fail (SF) or server signal degraded (SD). The SF/SD signal is forwarded by the server layer adaptation function A to the sub-network connection function.

SNC/I provides support for the (1 + 1) and (1:1) protection architectures.

7.3.2.2 *Non-intrusive Monitoring* The non-intrusive monitored SNCP scheme is generally referred to as SNC/N. In SNC/S the original trail termination function TT in the server layer network has no capability to monitor the performance of the server layer. For the monitoring of the (sub-)network connection listen-only trail termination functions TT_M are used. They should be available in the client layer and are indicated by the shaded area in Figure 7-22.

These TT_M are permanently connected to the server signal received from the server adaptation function A (indicated by the solid black line). A detected server signal failure SF or server signal degraded SD will be reported to the client layer connection function to initiate a protection switch.

SNC/N provides support for the (1 + 1) protection architecture. If the protection is uni-directional the APS protocol is not required.

7.3.2.3 *Sub-layer Trail Monitoring* The sub-layer trail monitoring SNCP in general referred to as SNC/S, can be modeled by expanding the sub-network connection points (CP) and inserting a protection sub-layer. The introduction of this sub-layer results in a trail protection of the working and protecting sub-layer trails, the inserted protection sub-layer is shown in the shaded area in Figure 7-23.

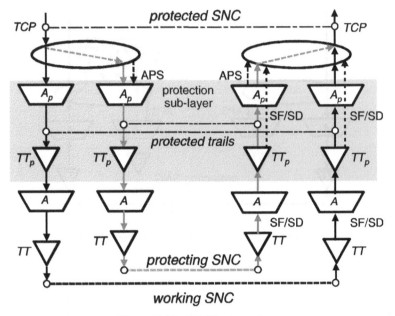

Figure 7-23. SNC/S protection

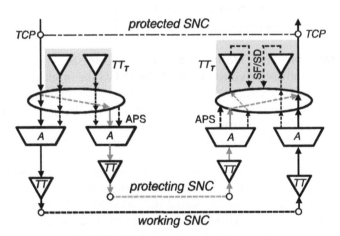

Figure 7-24. SNC/T protection

In the inserted sub-layer the protection trail termination function TTp provides the monitoring of the performance of the server signal. The monitoring results in a server signal fail (SF) or server signal degraded (SD) signal that is forwarded by the protection adaptation function Ap to the client layer connection function that will execute the protection switch.

SNC/S provides support for the $(1 + 1)$, $(1:1)$, $(1:n)$, $(m:n)$ and $(1:1)^n$ protection architectures.

7.3.2.4 *Test Monitoring* SNCP with Test Monitoring, SNC/T, is achieved by using a test signal that will be transported together with the normal traffic. As Figure 7-24 shows, the test signal generated in the test termination source function TT_T is multiplexed with both working and protecting entities at the source side. The test signal is extracted at the sink side and connected to the test termination sink function TT_T to determine the server layer performance.

SNC/T provides support for the $(1 + 1)$ and $(1:1)$ protection architectures.

CHAPTER 8

MAPPING METHODOLOGIES: FITTING THE PAYLOAD IN THE CONTAINER

In telecommunication transport networks are designed to carry client signals of different technologies. Initially the SDH/SONET networks carried PDH signals and later packet-based signals are carried as well. The OTN network initially carried SDH signals and other CBR signals and now is capable of carrying packet based signals too. The Packet Transport Network (PTN) is designed for the transport of packet based signals, e.g., Ethernet packets, but it is now able to transport SDH/SONET containers as well. For the transport through the network the client signal, also referred to as payload, is mapped into the technology specific containers, i.e., the C-n and C-n-X for SDH/SONET, the OPUk and OPUk-Xv for the OTN, or the payload area of the PTN traffic unit. The mapping process is part of the adaptation function.

8.1 SDH/SONET—CONTAINER C-n MAPPING

The SDH container used for mapping client signals is the C-n (n = 4, 3, 2, 12, 11). This container can be concatenated to provide more flexible bandwidth: C-n-X. The frame structure of these containers is described in Section 3.1.6 of Chapter 3. [ITU-T Rec. G707] describes several mapping methodologies. The following sections provide an overview of such mapping methodologies. Each mapping methodology is identified by the path signal label located in the C2 or K4 byte.

The ComSoc Guide to Next Generation Optical Transport: SDH/SONET/OTN,
by Huub van Helvoort
Copyright © 2009 Institute of Electrical and Electronics Engineers

8.1.1 Bit A-synchronous Mapping

This is the most common mapping methodology in SDH/SONET. All PDH signals can be mapped using this method, i.e., mapping E4 into VC-4, DS3 and E3 into VC3, DS2 into VC2, E1 into VC-12, and DS1 into VC-11. The PDH signals are regarded as CBR signals without a framing structure. The signals are mapped bit by bit. The bit-rate of the client signal is matched to the bit-rate of the SDH/SONET container by inserting fixed stuff bits. Rate adaptation is performed by bit stuffing as described in Section 6.2 of Chapter 6. Figure 8-1 shows an example of bit a-synchronous mapping. The mapping of E4 is indicated by C2 value "0x12," the mapping of DS3 and E3 is indicated by C2 value "0x04," the mapping of DS2, E1 and DS1 is indicated by K4 value "010."

8.1.2 Byte Synchronous Mapping

This mapping methodology is used when the tributary signal frame structure is recovered. Only the mapping of E1 into VC-12 and DS1 into VC-11 is defined. Before the PDH signals can be mapped their frame structure has to be recovered. The signals are mapped octet by octet. The container C-n clock is locked to the PDH recovered clock. Rate adaptation is performed by pointer justification as described in Section 6.1 of Chapter 6. The byte synchronous

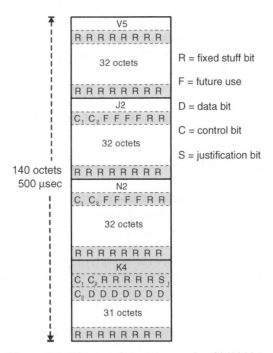

Figure 8-1. Bit a-synchronous mapping 2048 kbit/s

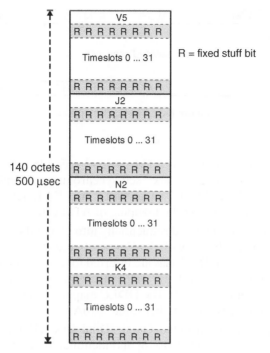

Figure 8-2. Byte synchronous mapping 2048 kbit/s

mapping is indicated by K4 value "100." Figure 8-2 shows an example of byte synchronous mapping.

8.1.3 Bit Synchronous Mapping

This mapping methodology can be used when the tributary signal clock is locked to the C-n clock. The mappings of DS2 into VC-2 {VT6} and DS1 into VC-11 {VT1.5} are defined but hardly used. In this case C_1 is set to 1, C_2 is set to 0, S_1 is unused and S_2 contains a data bit. The bit-synchronous mapping is indicated by K4 value "011."

8.1.4 Packet Mapping

Initially packet structured signals were mapped directly into SDH/SONET containers. Currently GFP mapping is used for all packet structured signals.

- ATM cells are mapped sequentially into C-n-X (n = 4 and 2) and C-n (n = 4, 3, 2, 12, and 11). The matching of the ATM bit-rate to the C-n or C-n_X bitrate is performed by the ATM cell creation process. ATM cell octets are aligned with the C-n octets and ATM cells are allowed to cross

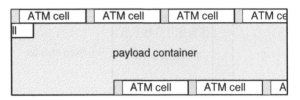

Figure 8-3. ATM cell mapping

C-n frame boundaries. The ATM payload area (48 bytes) is scrambled before mapping using a $1 + x^{43}$ self-synchronizing scrambler. The ATM mapping is indicated by C2 value "0x13" and K4 Extended Label value "0x09." Figure 8-3 shows the mapping of ATM cells.

- HDLC frames are mapped sequentially into C-n and C-n-X. The HDLC frames are octet aligned with the C-n octets and are allowed to cross C-n frame boundaries. HDLC flags are inserted for rate adaptation. The whole C-n is scrambled using a $1 + x^{43}$ self-synchronizing scrambler (scrambling a C-2/12/11 is not required). The HDLC maping is identified by C2 value "0x16" and K4 Extended Label value "0x0A."

- The mapping of Distributed Queue Dual Bus (DQDB) into a C-4 is described in [ETSI ETS 300 216] and identified by C2 value "0x14."

- Fibre Distributed Data Interface (FDDI) mapping into C-4 is described in [ITU-T Rec. G.707] and identified by C2 value "0x15."

- A 10 Gbit/s Ethernet, 10G base-W or WAN PHY, signal can be mapped into a VC-4-64c as defined in [IEEE 802.3ae] sections 49 and 50. The 64B/66B coded signal is bit synchronous mapped independent of the Ethernet block and packet boundaries. The path signal label C2 is set to "0x1A." This mapping is also defined in [ITU-T Rec. G.707] Annex F.

- Other packet-based signals are mapped using GFP as described in Section 7.3. The GFP frame octets are aligned with the C-n octets and GFP frames are allowed to cross C-n frame boundaries. The GFP frames can be mapped in any C-n or C-n-X and are identified by C2 value "0x1B" and K4 Extended Label value "0x0D."

8.1.5 Special Mappings

In case separate OTN domains have to be connected via an SDH network an ODUk can be mapped into a C-4-X. The mapping is byte a-synchronous, which means that the ODUk and the C-4-X are octet aligned and rate justification uses an octet for stuffing. The octet stuffing is similar to the bit stuffing described in Section 6.2 of Chapter 6. In this case five bits are used for justification control and a single octet is used for negative justification. The ODU1 is mapped into a C-4-17 and the ODU2 is mapped into a C-4-68. The mapping is identified by C2 value "0x20."

8.2   OTN—CONTAINER OPUk MAPPING

The OTN container used for mapping client signals is the OPUk ($k = 1, 2, 3$). This container can be virtually concatenated to provide more flexible bandwidth: OPUk-Xv. The frame structure of these containers is described in Section 3.2.3 of Chapter 3. [ITU-T Rec. G709] describes several mapping methodologies.

The following sections provide an overview. Each mapping methodology is identified by the path signal label located in the C2 or K4 byte.

Note: *Currently (2009) the ITU-T is developing an extension of the OTN hierarchy by defining an ODU4 to facilitate the transport of 40 Gbit/s and 100 Gbit/s Ethernet signals. A Generic Mapping Procedure (GMP) to provide bit-rate agnostic mapping is developed at the same time.*

8.2.1   CBR and STM Mapping

The OTN structure was designed to transport CBR signals. At that time the SDH/SONET frame structures STM-16 {OC-48}, STM-64 {OC-192} and STM-256 {OC-768} were the candidates considered for transport in the OTN. The size of the OPUk is related to these signals and designed to transport the CBRn signals with bitrates $4^{(n-1)} \times 2488.32$ Mbit/s (n = 1, 2 and 3), generally referred to as CBR2G5, CBR10G and CBR40G. The CBR signals are mapped bit by bit into the OPUk frame structure.

A CBR2G5 signal, e.g., an STM-16, is mapped into an OPU1 as shown in Figure 8-4.

A CBR10G signal, e.g., an STM-64, is mapped into an OPU2 as shown in Figure 8-5. Sixteen octet columns containing fixed stuff, located in columns 1905 ... 1920 of the OPU2, are present to match the bandwidth of the CBR10G signal.

CBR40G signal, e.g., an STM-256, is mapped into an OPU2 as shown in Figure 8-6. Two times 16 octet columns containing fixed stuff, located in columns 1265 ... 1280 and 2545 ... 2560 of the OPU3, are present to match the bandwidth of the CBR40G signal.

The mapping of the CBR signals can be bit a-synchronous where the clock of the CBR signal is independent of the OPUk clock. In this case the positive

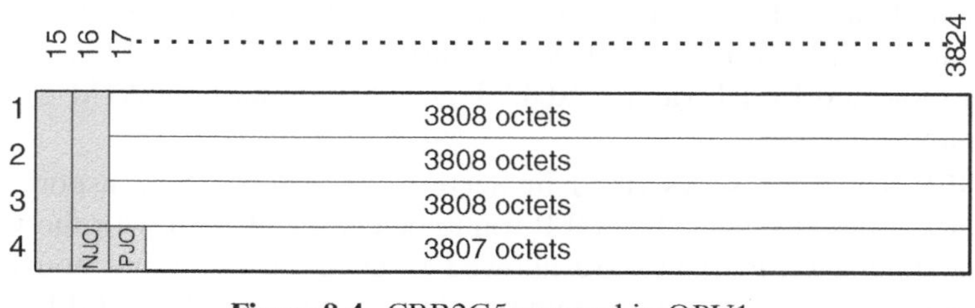

Figure 8-4. CBR2G5 mapped in OPU1

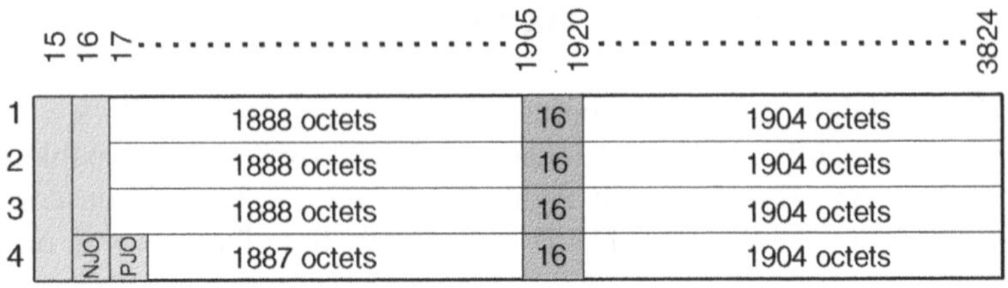

	15 16 17	....	1905	1920	....	3824
1		1888 octets	16	1904 octets		
2		1888 octets	16	1904 octets		
3		1888 octets	16	1904 octets		
4	NJO PJO	1887 octets	16	1904 octets		

Figure 8-5. CBR10G mapped in OPU2

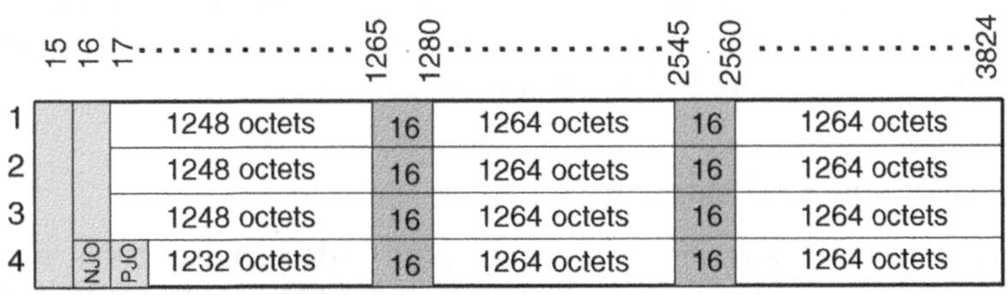

	15 16 17	....	1265	1280	....	2545	2560	....	3824
1		1248 octets	16	1264 octets	16	1264 octets			
2		1248 octets	16	1264 octets	16	1264 octets			
3		1248 octets	16	1264 octets	16	1264 octets			
4	NJO PJO	1232 octets	16	1264 octets	16	1264 octets			

Figure 8-6. CBR40G mapped in OPU3

and negative bit-rate justification described in Section 6.3 of Chapter 6 is used. This is indicated by setting the OPUk Payload Type (PT) value to "0x02."

If the OPUk clock is locked to the CBR clock the mapping is synchronous and no justification is required. That is, the PJO octet always contains data and the NJO octet is not used. This mapping is indicated by setting the value of PT to "0x03."

8.2.2 Packet Mapping

Currently ATM mapping and GFP mapping are defined for the OTN.

- ATM cells are mapped sequentially into an OPUk or OPUk-Xv. The matching of the ATM bit-rate to the OPUk or OPUk-Xv bit-rate is performed by the ATM cell creation process. ATM cell octets are aligned with the OPUk octets and ATM cells are allowed to cross OPUk frame boundaries. The ATM payload area (48 bytes) is scrambled before mapping using a $1 + x^{43}$ self-synchronising scrambler. The ATM mapping is indicated by PT value "0x04." Figure 8-3 shows how ATM cells are mapped.
- GFP mapping as described in Section 7.3 is used to transport other packet-based signals. The GFP frame octets are aligned with the OPUk octets and GFP frames are allowed to cross OPUk frame boundaries similar to the ATM cell mapping depicted in Figure 8-3. The GFP frames

can be mapped in any OPUk or OPUk-Xv and are identified by PT value "0x05." A 10 Gbit/s Ethernet, 10G base-R (LAN PHY), signal can be mapped using GFP-F.

- A 10 Gbit/s Ethernet, 10G base-W (WAN PHY), signal can be mapped into an ODU2 using the CBR10G mapping described above.

8.2.3   Special Mappings

Transporting a 10G base-R (LAN PHY) Ethernet signal bit transparently, i.e., including preamble, start of frame delineation, and interpacket gap, by mapping it into an OPU2 is not possible without increasing the clock-rate of the OPU2. The resulting OUP2e is nonstandard and is described in the non-normative recommendation [ITU-T Rec. G.Sup43].

8.2.4   Future Mappings

At the time this book is published the IEEE Task Force 802.3 ba is defining the standard for *40 Gigabit Ethernet* (40 GbE) and *100 Gigabit Ethernet* (100 GbE). The ITU-T experts will then extend the OTN hierarchy to be able to transport these new signals. The ODU3 may have to be adapted and the new OUD4 needs to be defined. They are also considering a new mapping methodology: *Generic Mapping Procedure* (GMP), which will provide more flexibility for mapping client signals including the 40 GbE and 100 GbE.

8.3   GFP MAPPING

To avoid a proliferation of mapping methodologies for packet-based signals the ITU-T has defined a *Generic Framing Procedure* (GFP): [ITU-T Rec. G.7042]. The definition includes the frame formats of the GFP *Protocol Data Units* (PDUs) and the procedure for mapping the client signals into the GFP PDUs.

Figure 8-7 illustrates the relationship between the client layer packet based signals, GFP mapping, and the server layer *Constant Bit-Rate* (CBR) signals.

The client specific aspects of GFP concern the mapping of the different packet structures into a GFP frame. GFP can be used for:

- The encapsulation and transport of entire client frames: *Frame mapped GFP* (GFP-F): a single client PDU, e.g., an Ethernet MAC frame or an IP/PPP frame is mapped into a single GFP-F PDU.
- The transport of blocks of client data characters: *Transparent mapped GFP* (GFP-T): a number of client data characters, e.g., Fibre Channel or ESCON 8B/10B block codes, are mapped into a single GFP-T PDU.

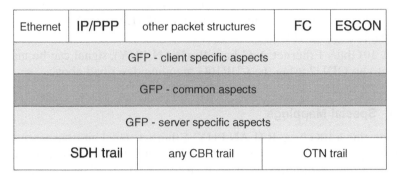

Figure 8-7. GFP between client layer and server layer

The details of these mappings are described in sections 7.3.2 and 7.3.3.

The common aspects of GFP concern the GFP frame structure used for client data transport, client management frames, and GFP idle frames. Details of GFP frame structures are provided in Section 7.3.1.

The server-specific aspects of GFP concern the mapping of GFP frames into the server structures. The mapping into an SDH VC-*n* is specified in [ITU-T Rec. G.707]. The mapping into an OTN ODUk payload is specified in [ITU-T Rec. G.709]. Generally the bandwidth of the server signal should be larger than that of the client signal. GFP idle frames are used when insufficient client data is available.

8.3.1   GFP Frame Structure

The GFP frame structure consists of a four octet Core Header and a 0...65535 octets Payload Area. The Core Header consists of a two octet Payload Length Indication and a two octet *core Header Error Check* (cHEC). With the cHEC code single bit errors are corrected. The Payload Area consists of a Payload Header, a client payload area, and optionally a four octet GFP *Frame Check Sequence* (FCS). The Payload Header consists of a Type Header and optionally an Extension Header. The Type Header consists of a two octet type field containing a three bit *Payload Type Indicator* (PTI), a single bit Payload FCS Indicator (PFI, 0 = no GFP FCS), a four bit Extension Header Identifier (EXI, 0 = no extension header), and an eight bit *User Payload Identifier* (UPI), and a two octet *Type HEC* (tHEC). With the tHEC code single bit errors are corrected. The Extension Header consists of up to 58 octets with additional payload information and a two octet *Extension HEC* (eHEC). With the eHEC code single bit errors may be corrected. Figure 8-8 shows the GFP frame structures.

Note that the GFP idle frame consists of four octets (PLI = "0x0000" and cHEC = "0x0000"). All Core headers are scrambled using the Barker-like

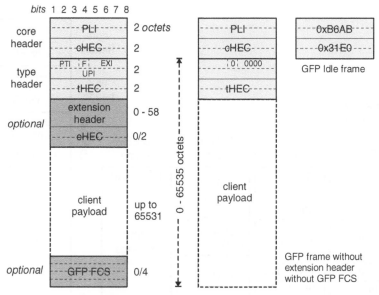

Figure 8-8. GFP frame structures

sequence shown in Figure 8-8 for the GFP idle frame. The Payload Area of the GFP frame is scrambled using a $1 + x^{43}$ self-synchronizing scrambler.

Currently the PTI value "000" indicates "client data" and the associated UPI identifies the client signal and its mapping methodology, e.g., "0x01" = Frame mapped Ethernet, "0x05" = Transparent mapped ESCON. The PTI value "100" indicates "client management" and the associated UPI is used to transfer OAM information, e.g., "0x01" = Loss of Client signal, "0x05" = Remote Defect Indication.

8.3.2   Frame Mapped GFP (GFP-F)

GFP-F transfers client data frame by frame; each client frame is mapped into the client payload area after discarding possible preamble or flags. Figure 8-9 shows an example of several client signals. The optional GFP FCS is calculated over the client payload area only.

8.3.3   Transparent Mapped GFP (GFP-T)

GFP-T transfers client data that are 8B/10B block-coded and require low transfer latency. Instead of buffering a whole frame, like GFP-F, the individual characters of the client signal are mapped into fixed length GFP frames. The mapping is independent of the character content, data or control. Figure 8-10 shows an example of several client signals.

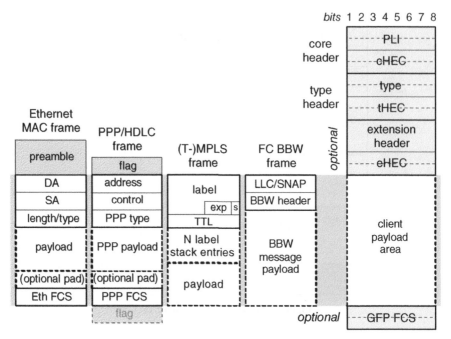

Figure 8-9. GFP-F mapping

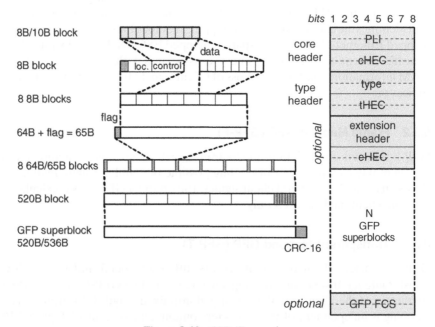

Figure 8-10. GFP-T mapping

The following steps are taken by the mapping process:

- The 8B/10B characters are decoded into data octets or control characters. Because there are only 12 control codes they can be identified by a four-bit code.
- Data and control are mapped into eight octet blocks. If there are one or more control characters in this block they are moved to the beginning of the block together with a three-bit coded location code that is used to insert the control characters in their original sequence it the character stream. A flag bit is used to indicate the last coded control character in the block.
- A flag bit is added to the block to indicate that the 64B/65B block contains control characters (flag = 0 data characters only).
- Eight successive 64B/65B blocks are grouped together to form a 520B block where the flag bits of the individual 64B/65B blocks are collected into the last octet of the 520B block.
- A GFP superblock is formed by calculating a 18 bit CRC over the 520B block and inserting the result in two additional octets of the 520B/536B superblock.
- N GFP superblocks are mapped into the client payload area of a GFP frame. The value of N depends on the client data-rate and the server data-rate. Appendix IV of [ITU-T Rec. G.7041] contains a table with suggested values of N.

For data rate adaptation 65B_PAD codes are inserted.

The GFP mapping process can start the transfer as soon as the first 64B/65B block in the superblock has been formed.

CHAPTER 9

CONCATENATION: TO GROW AND PROVIDE FLEXIBILITY

The first generation of SDH and SONET equipment supported the transport and switching of the basic frame structure: the VC-4 or {STS-1}. This was sufficient to transport the highest order PDH multiplexed signals present at that time. When the definitions of SDH and SONET were aligned by using the same frame structures great care was taken that the SDH network would be able to transport {STS-1 SPE} frames and that the SONET network would be able to transport VC-4 frames. The former is achieved by defining equivalent frame structures for the VC-3 and the {STS-1 SPE}. The latter is achieved by concatenating three {STS-1 SPE} frame structures such that the resulting frame structure is equivalent to that of a VC-4. When the SDH and SONET increased the multiplexing capabilities the concatenation principle was used to provide containers with a higher bandwidth capacity to support the transport of high speed data signals.

More applications of concatenation are described in the following sections.

9.1 CONTIGUOUS CONCATENATION (CCAT)

Apart from the alignment of SDH and SONET, frame structures concatenation is also used to provide payload containers that have a larger capacity than

The ComSoc Guide to Next Generation Optical Transport: SDH/SONET/OTN,
by Huub van Helvoort

the basic VC-4 or {STS-1 SPE}. This is achieved by multiplying the number of columns in the frame structure of a container C-n by an integer value X providing a container C-n-X. This container can be transported by a dedicated virtual container with a continuous payload area: the VC-n-Xc a concatenation of X VC-n. This is the contiguous concatenated virtual container. Because the ITU-T defines each next higher order multiplex with a size four times larger than the existing multiplex and *Contiguous conCATenation* (CCAT) was at first used by higher order multiplexes, the value of X is limited to orders of four, i.e., X = 4, 16, 64, or 256. Originally SONET allowed any value of X in {STS-1-Xc SPE}, however, for the interworking with SDH equipment contiguous concatenation is restricted to {STS-Nc SPE} with N = 3 * X (X = 1, 4, 16, 64, 256).

In Chapter 3, Figure 3-16 shows the frame structure of a VC-4-Xc that contains a C-4-X container. Figure 9-1 shows the mapping of the C-4-X container in a VC-4-Xc and POH bytes processed by the adaptation and termination functions.

The VC–4–Xc is constructed by byte interleaving the X individual VC–4s. The result is that the columns containing the *Path Overhead* (POH) of the individual VC–4s are located in the first X columns of the VC–4–Xc. Out of the X columns containing the POH bytes from the original VC–4s only one, located in the first column, is used as the POH common for the whole VC–4–Xc and consists of nine bytes/octets. Columns 2 to X inclusive contain fixed stuff bytes. The remaining X * 260 columns provide the payload area of the VC–4–Xc that has exactly the same size as the C–4–X.

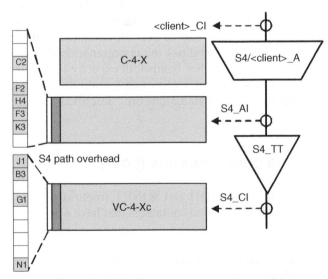

Figure 9-1. Contiguous concatenation model

The VC–4–Xc is transported in X contiguous AU–4s in an STM–N signal. The first column of the VC–4–Xc will always be located in the first AU–4 of X contiguous AU–4s. For example, for a VC–4–4c transported in an STM–16 this will be either the AU–4 #1, AU–4 #5, AU–4 #9, or AU–4 #13. The pointer to this first AU–4 indicates the position of the first octet in the POH of the VC–4–Xc, i.e., the octet in [1, 1] of the frame. The pointers to the remaining AU–4s, i.e., AU–4 #2 to #X, contain the concatenation indication: "1001xx11 11111111." See also Table 6-2 in Chapter 6.

The only contiguous concatenated lower order container is the C-2-X, which is transported by a VC–2–Xc (X = 1 ... 7).

9.2 VIRTUAL CONCATENATION (VCAT)

When contiguous concatenation of VC-4s was defined the installed equipment in the SDH network supported cross-connection of nonconcatenated signals only, i.e., in the network only payload containers up to container size C-4 could be transported. However, even when the equipment supports contiguous concatenated signals it is still possible that a VC-4-Xc cannot be provisioned because a set of X consecutive VC-4s is not available *and* the first VC-4 of the set cannot be mapped into an AU-4 numbered (1 + nX, n = 0, 1, 2, ...). See Figure 9-2 for an example where a VC-4-4c needs to be provisioned in an STM-16. One VC-4-4c is already provisioned in AU-4 #1 ... #4, and two VC-4s are provisioned in AU-4 #6 and AU-4 #11. There are three opportunities to provision a set of four consecutive VC-4s: starting at AU-4 #7, #12, and #13. However, only the set starting at AU-4 #13 can be used to provision a VC-4-4c. To provision another VC-4-4c at least one of the VC-4 has to be moved to another AU-4, this move will cause a hit in the payload of the concerned VC-4.

To overcome these limitations of contiguous concatenation *Virtual con-CATenation* (VCAT) has been defined.

The payload container of a virtual concatenated signal is still a contiguous container of size C–n–X, but by VCAT it is transported through the network as X individual VC-ns each with a payload container size C–n. The X VC-ns, i.e., the VC–n–Xv, is in general referred to as a *Virtual Concatenation Group* (VCG), the X individual VC-ns in the VCG are referred to as members of the VCG. The VCAT functionality is required only in the path terminating equipment; the intermediate nodes are not aware of the fact that a transported VC-n

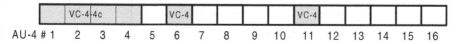

Figure 9-2. VC-4-4c provisioning in an STM-16

is part of a VCG. For backwards compatibility the ITU-T has defined an interworking function that provides a conversion between a CCAT VC–n–Xc and a VCAT VC–n–Xv.

9.2.1 Payload Distribution and Reconstruction

The distribution of the content of a C–n–X over the X individual containers C-n is column-wise: this is shown in the VC-n-Xv structure in Figure 3-19 in Chapter 3. Figure 9-3 shows the mapping of the C-n-X container in a VC-n-Xv. This figure also shows where the VCAT overhead is inserted and extracted. Note that the LCAS specific overhead is grayed.

The VC-n POH is not shown in Figure 9-3; it is shown in Figures 4-2 and 4-19 of Chapter 4. The HO VCAT uses the H4 byte and LO VCAT uses the K4 bit 2.

OTN supports VCAT in a similar way; X OPUk can transport an OPUk-X payload. The OPUk-Xv frame structure is shown in Figure 3-24 in Chapter 3.

To recognize the order of mapping the C-n-X into the X VC-ns of a VCG, each VC-n is assigned a unique *SeQuence number* (SQ): The first VC-n has the SQ value "0," the second "1," and the last VC-n has the SQ value "(X-1)."

Because each VC-n that is a member of a VCG is transported individually through the network between the VCAT termination functions it can follow a network path that is diverse from the path followed by other member VC-n in the VCG.

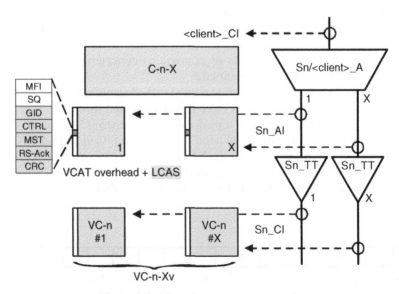

Figure 9-3. Virtual concatenation model

Due to difference in the physical length of the path followed by each member in the VCG the physical delay of each member may be different. Also the number of intermediate nodes in each path may be different and due to the transfer delay of a node this will also introduce different delays for each path. The total of physical delay and transfer delay for each VC–n, i.e., the propagation delay, may be different between the VC–ns in a VCG. The difference in delay between the fastest and the slowest VC–n in the VCG is called the *differential delay*. This differential delay has to be measured at the sink side path termination function to be able to realign the member VC-ns before the original C-n-X can be reconstructed. The delay compensation has to be dynamic because it is possible that the propagation delay changes due to a protection switch in a server layer, which changes the length of the path. To measure the differential delay the *Multi-Frame Indicator* (MFI) has been defined. Each frame of the original C–n–X has a unique MFI value, i.e., the value of a counter that is incremented each frame. When the content of the C-n-X frame is copied to the individual C-n frames, the value of the MFI is also copied. By comparing the MFI values at the receive side the differential delay can be determined. The member with the lowest delay will have relatively the lowest MFI value; the member with the largest delay will have the highest MFI value. Compensation of the differential delay is accomplished by delaying the faster members in the VCG. This is accomplished by using variable buffers for each member in a VCG. The maximum allowed differential delay is 256 ms. With a physical propagation delay of 5 ms per 1,000 km the maximum acceptable difference in physical path is 50,000 km, enough for global applications.

At the sink side the received C-ns are first aligned using the MFI, i.e., the differential delay is compensated. Then the C-n-X is reconstructed by taking the first column of each of the X C-n and ordering them using the SQ numbers, next the X second columns are taken, etc. until the X last columns are taken and the C-n-X frame is complete.

9.2.2   Higher Order VCAT Overhead

For higher order VCAT (VC-4, VC-3, {STS-1 SPE}, {STS-3c SPE}) 16 consecutive H4 bytes are used to form a VCAT OH multi-frame, i.e., the *control packet*. The multi-framing is provided by the bits [5...8] of each H4 byte as shown in Table 9-1. These four bits are also used as the four *Least Significant Bits* (LSB) of the MFI and identified by MFI-1. The eight *Most Significant Bits* (MSB) of the MFI are identified by MFI-2. The value of MFI–2 is incremented every VCAT multi-frame and counts from 0 to 255.

Table 9-1 shows the location of the MFI-1, the MFI-2 and the SQ fields in the HO VCAT multi-frame. The fields marked *Reserved* are set to "0."

Virtual concatenation for SDH is described in [ITU-T Rec. G.707]. Defects related to VCAT are described in [ITU-T Rec. G.806]. VCAT for SONET is described in [ANSI T1.105].

TABLE 9-1. HO VCAT multi-frame using H4 byte

Bit 1	Bit 2	Bit 3	Bit 4	Bit 5	Bit 6	Bit 7	Bit 8	1st multi-frame number	2nd multi-frame number
H4 byte									
				MFI-1 bits [1 ... 4]					
Reserved				1	1	0	1	13	
SQ				1	1	1	0	14	n–1
				1	1	1	1	15	
MFI-2				0	0	0	0	0	
				0	0	0	1	1	
				0	0	1	0	2	
				0	0	1	1	3	
				0	1	0	0	4	
				0	1	0	1	5	
				0	1	1	0	6	
Reserved				0	1	1	1	7	n
				1	0	0	0	8	
				1	0	0	1	9	
				1	0	1	0	10	
				1	0	1	1	11	
				1	1	0	0	12	
				1	1	0	1	13	
SQ				1	1	1	0	14	
				1	1	1	1	15	
MFI-2				0	0	0	0	0	
				0	0	0	1	1	n + 1
Reserved				0	0	1	0	2	

9.2.3 Lower Order VCAT Overhead

In the existing VC-m POH there were not enough reserved bits available to be used for the overhead of lower order VCAT (VC-2, VC-12, VC-11, {VT6 SPE}, {VT2 SPE}, {VT1.5 SPE}). To provide the required bits a two stage process was defined:

Stage 1: The value of the V5 byte bits [5, 6, 7] is set to "101," i.e., "extended signal label." In the K4 byte bit 1 a 32 bit string is inserted as shown in Table 9-2. This 32 bit string contains the Multi-Frame Alignment Signal, MFAS, an eight bit extended signal label field, a separator bit with value "0" and 12 bits reserved for future use all set to "0," it is repeated every 16 ms.

Stage 2: K4 byte bit 2 which is aligned with the multi-frame in K4 byte bit 1 is used to transport the lower order VCAT OH multi-frame.

TABLE 9-2. VCAT + LCAS parameters

Fieldname	Remarks
MFI	Multi-Frame Indication—used for determining the differential delay • 12 bits in HO VCAT • 5 bits in LO VCAT • 16 bits for OTN VCAT
SQ	SeQuence number—used to realign the columns when reconstructing the original payload • 8 bit for HO VCAT, allowing up to 256 members • 6 bit for LO VCAT, allowing up to 64 members • 8 bit for OTN VCAT, allowing up to 256 members In *non-LCAS mode*, i.e., VCAT only, the set of SQ numbers is provisioned by the EMF at both the Source side and the Sink side of the trail. Changing the size of a VCG can only be accomplished by completely removal of the original VCG and setting up a new VCG between the trail termination functions. In *LCAS mode* the assignment of the SQ numbers is performed by the LCAS procedure.
MST *(LCAS)*	Member STatus—used to transfer the status (OK or FAIL) of each member at the Sink towards the Source side. 1 bit per member, 8 bits per control packet, it is multi-framed using the MFI to provide 256 bits in HO VCAT and 64 bits in LO VCAT
RS-Ack *(LCAS)*	Re-Sequence Acknowledge—used to inform the Source side that the Sink has detected a change in the SQ numbering, or a change in the number of members in the VCG
CTRL *(LCAS)*	ConTRoL code—used to transfer the status of the member at the Source towards the Sink side. The following states exist: • FIXED—to indicate non-LCAS mode of operation • ADD—the member is about to be added to the VCG • NORM—normal operation state, the member carries payload • EOS—normal operation state, the member carries payload and this member has the highest SQ number • IDLE—the member is not active in the VCG and can be removed • DNU—indicates that the payload of the member shall not be used because a network fault has been detected at the Sink side
GID *(LCAS)*	Group IDentity—may be used to identify a VCG, all members in a VCG carry the same GID
CRC *(LCAS)*	Cyclic Redundancy Check—used to validate the control packet. If the CRC fails the control packet is discarded. In OTN VCAT + LCAS the CRC in VCOH3 is calculated over the preceding VCOH1 and VCOH2.
Reserved or *R*	Reserved bits—used for future extensions, set to "0," ignored at reception

TABLE 9-3. HOVCAT + LCAS multi-frame using H4 byte

H4 byte								1st multi-frame number	2nd multi-frame number
Bit 1	Bit 2	Bit 3	Bit 4	Bit 5	Bit 6	Bit 7	Bit 8		
				MFI-1					
Reserved				0	1	0	1	5	
CRC-8				0	1	1	0	6	
				0	1	1	1	7	
MST				1	0	0	0	8	
				1	0	0	1	9	
R	R	R	RS-Ack	1	0	1	0	10	n
Reserved				1	0	1	1	11	
				1	1	0	0	12	
				1	1	0	1	13	
SQ				1	1	1	0	14	
				1	1	1	1	15	
MFI-2				0	0	0	0	0	
				0	0	0	1	1	
CTRL				0	0	1	0	2	
R	R	R	GID	0	0	1	1	3	
Reserved				0	1	0	0	4	
				0	1	0	1	5	n + 1
CRC-8 LSB				0	1	1	0	6	
				0	1	1	1	7	
MST				1	0	0	0	8	
				1	0	0	1	9	
R	R	R	RS-Ack	1	0	1	0	10	

The LO VCAT multi-frame is shown in Table 9-4, for VCAT only, the CTRL, GID, MST and RSA fields are set to "0."

9.2.4 OTN VCAT Overhead

The OTN VCAT (OPUk-Xv) information is located in [1, 2, 3 16] and it is identified by the bytes VCOH1, VCOH2, and VCOH3. The five LSB of the MFAS in the frame alignment overhead [1 7] of the ODUk are used to provide a 32 word VCAT multi-frame. The frame is shown in Table 9-5, for VCAT only, the CTRL, GID, MST and RSA fields are set to "0".

Virtual concatenation for OTN is described in [ITU-T Rec. G.709].

9.3 LINK CAPACITY ADJUSTMENT SCHEME (LCAS)

Because it was not possible to change the size of an existing VCG, i.e., change the number of members, without having to delete the VCG and create a new

TABLE 9-4. LO VCAT multi frame using K4 bit 2

MFAS in K4 bit 1		VCAT OH in K4 bit 2		Bit number
Function	Value	Function	Value	
	0	MFI	MSB	1
	1			2
	1			3
	1			4
	1		LSB	5
Multi-Frame Alignment Signal	1	SQ	MSB	6
	1			7
	1			8
	1			9
	1			10
	0		LSB	11
		CTRL		12
				13
Extended signal label				14
				15
Not used by VCAT		GID		16
		Reserved	0	17
			0	18
			0	19
Fixed	0		0	20
	0	RS-Ack		21
	0			22
	0			23
	0			24
	0	MST		25
Reserved	0			26
	0			27
	0			28
	0			29
	0			30
	0	CRC-3		31
	0			32

VCG, a procedure was developed to enable the changes without affecting the transported client information. This procedure is specified as the *Link Capacity Adjustment Scheme* (LCAS). [Comment: It would be useful if you can add a brief history or rationale of introducing LCAS.]

To support LCAS a number of additional parameters are required. These are described in Table 9-2. The table includes a description of the MFI and SQ fields that are also used in non-LCAS mode.

The allocation of the LCAS parameters in the VCAT multi-frame is shown for HO VCAT in Table 9-3. The LCAS protocol and associated parameters

TABLE 9-5. OTN VCAT + LCAS multi-frame using VCOH bytes

VCOH1	VCOH2 MST bits								VCOH3	MFAS LSB
MFI-1	0	1	2	3	4	5	6	7	CRC-8	0
MFI-2	8	9							CRC-8	1
Reserved									CRC-8	2
									CRC-8	3
SQ									CRC-8	4
CTRL GID RSA R R									CRC-8	5
									CRC-8	6
									CRC-8	7
Reserved									CRC-8	:
									CRC-8	30
	249	250	251	252	253	254	255		CRC-8	31

has been developed after the HO VCAT multi-frame was defined and because it is required that the CRC is located at the end of the control packet, the HO VCAT + LCAS multi-frame starts at MFI-1 value 8 and ends at MFI-1 value 7. The LO VCAT + LCAS multi-frame is shown in Table 9-4. The OTN VCAT + LCAS multi-frame is depicted in Table 9-5.

The generic Link Capacity Adjustment Scheme (LCAS) protocol is described in [ITU-T Rec. G.7042].

An extensive description of VCAT and LCAS can be found in the books:

- *Next Generation SDH/SONET: Evolution or Revolution*, John Wiley & Sons, ISBN 0-470-09120-7.
- *Optical Networking Standards: A Comprehensive Guide for Professionals*, Chapter 4, Springer, ISBN 978-0-387-24062-6.

Short descriptions of VCAT can be found in the following magazines:

- *Lightwave* 12/2001: *"Standards bring more flexibility to SONET/SDH."*
- *Optical Networks* 01-02/2003: *"Standards bring more flexibility to optical transport."*
- *IEEE Communications* 05/2006: *"VCAT/LCAS in a clamshell."*

CHAPTER 10

SDH VS. SONET OVERHEAD PROCESSING: COMMONALITIES AND DIFFERENCES

Even though the frame structures of SDH and SONET are exactly the same, including the allocation of the overhead bits and bytes, the processing of the overhead bytes can be different due to the difference in *Operation, Administration, and Management* (OAM) philosophies of the standards organizations that developed SDH and SONET.

This chapter is dedicated to providing an overview of the overhead bits and bytes of SDH and SONET. It highlights the commonalities and differences of the overhead processing in these technologies. Tables 10-1 to 10-5 show the commonalities and differences per layer and per signal.

Some of the differences are also caused by the multiplexing hierarchy in SDH and SONET. As depicted in Figure 3-1 in Chapter 3, two levels of multiplexing are recognized, generally referred to as lower order and higher order multiplex. Because SDH is designed to transport PDH E4 and lower multiplexes, the VC-4 is considered to be the higher order signal and the VC-3, VC2, VC-12, and VC-11 the lower order signals multiplexed into the VC-4. SONET is designed to transport PDH DS3 and lower order multiplexes. The STS-1 SPE is considered to be the higher order signal and the VT6, VT2, and VT1.5 the lower order signals multiplexed into the STS-1 SPE. In SDH the VC-4 and its pointer (i.e., the AU-4) are mapped into AUG-N (N = 1, 4, 16, 64, 256) and in SONET the VC-3 {STS-1 SPE} and its pointer (i.e., the AU-3) are mapped into AUG-N as well. This means that SDH and SONET equipment

The ComSoc Guide to Next Generation Optical Transport: SDH/SONET/OTN,
by Huub van Helvoort

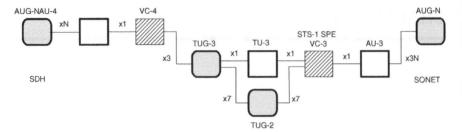

Figure 10-1. Interconnection re-multiplexing

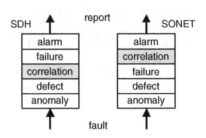

Figure 10-2. Correlation process

is interconnected a re-multiplexing is required. This is shown in Figure 10-1, which also contains the TUG-2 for the case where the VC-3 {STS-1 SPE} is sub-structured.

The objective for fault handling is "a single fault shall raise a single alarm." Differences exist because in SDH fault handling is independent of the connectivity in the network element, while in SONET fault handling depends on the connectivity in the network element. In SDH the *defects* are correlated, while in SONET the *failures* are correlated. See Figure 10-2 for the fault handling chain in SDH and SONET.

Similarly, in *performance monitoring* (PM), this is based on counting bit errors, transmission defects, pointer actions, and protection switching events in one second intervals. SDH PM is independent of the connectivity in the network element while SONET does not count connectivity defects. Normally PM is uni-directional, however SDH requires in addition bi-directional PM for QoS and billing.

Tables 10-1 to 10-5 also show the processing performed by *Non-Intrusive Monitors* (NIM).

In general, NIMs are used in the path layers for the following applications:

- to provide detection of connectivity and error defects for SNC/N protection switching

- to be a "work around" for tandem connection monitoring when TCM overhead processing (N1 bytes) is not supported
- to aid in fault localization within a VC {STS} trail
- to perform single-ended maintenance of a VC {STS} trail by monitoring at an intermediate network element.

Following is a list of the abbreviations used in Tables 10-1 to 10-5:

A	Adaptation function
AIS	Alarm Indication Signal
APS	Automatic Protection Switch
AU	Administrative Unit
BIP	Bit Interleaved Parity
C	Connection function
DEG	DEGraded (based on 10^{-x} algorithm)
DEG(EB)	DEGraded (based on Errored Blocks per second)
{ERDI}	Enhanced RDI
EXC	EXCessive based on 10^{-x} algorithm
F_EBC	Far end—Errored Block Count
F_DS	Far end—Defect Second
FOP	Failure Of Protocol
LOF	Loss Of Frame
LOP	Loss Of Pointer
LOPS	LOP Seconds
N_EBC	Near end—Errored Block Count
N_DS	Near end—Defect Second
NDF	New Data Flag
NIM	Non-Intrusive Monitor function
{NPJC}	Negative Pointer Justification Counter
{OFS}	Out of Frame Second
{PDI}	Payload Defect Indicator
{PJCDiff}	Pointer Justification Count Difference
{PJCS}	Pointer Justification Count Seconds
PJE–	negative Pointer Justification Event
PJE+	positive Pointer Justification Event
PLM	PayLoad Mismatch
PM	Performance Monitoring
{PPJC}	Positive Pointer Justification Counter
RDI	Remote Defect Indication

{RDIc}	Remote connectivity Defect Indication
{RDIp}	Remote payload Defect Indication
{RDIs}	Remote server Defect Indication
REI	Remote Error Indication
RFI	Remote Failure Indication
SD	Signal Degrade
{SEFS}	Severely Errored Frame Second
SF	Signal fail
{SPE}	Synchronous Payload Envelope
{STS}	Synchronous Transport Signal
{TCA}	Threshold Crossing Alert
TCM	Tandem Connection Monitor
TIM	Trace Identifier Mismatch
TT	Trail Termination function
TU	Tributary Unit
UNEQ	Unequipped
VC	Virtual Container
VCAT	Virtual conCATenation
{VT}	Virtual Tributary

If the overhead information is processed it is indicated by a "y." An "n" indicates not processed.

Differences in processing between SDH and SONET are indicated by shading.

The information in the tables is based on [ITU-T Rec. 707], [ITU-T Rec. G.783] and [ETSI EN 300 417-x-1 (x = 1, ..., 7, 9, 10)] for SDH and [ANSI T1.105], [ANSI T1.231] and [Telcordia GR253] for SONET.

OH byte	Direction	Process	Location	SDH Regenerator Section RS-0/1/4/16/64/256	SONET Section STS-1/3/12/48/192/768
A1A2—frame alignment signal	Source	Insertion	TT	y	y
	Sink	Defect detection—reporting	TT	OOF: 5 frames LOF: 3 ms with filtering	OOF: 4 frames LOF: 3 ms
		PM—{OFS [SEFS]}	TT	n	y
		PM—N_DS	TT	y	y
		Consequent action—AIS downstream	TT	y	y
		Consequent action—RDI upstream	TT	y	y
		Protection switching—SF	TT	y	y
B1—BIP8	Source	Calculation	TT	y	y
	Sink	Defect detection—reporting	TT	y—DEG (in regen. appl.)	n
		PM—N_EBC	TT	y, block size under study	y, based on BIP8 violations
		Protection Switching—SF [10e-x x = 3,4,5]	TT	n	n [x = 3]
		Protection Switching—SD [10e-x x = 5...9]	TT	n	n [x = 6]
		{TCA} 10e-x x = 5...9]	TT	n	y [x = 6]
C1—...	Source	Insertion	—	#1..#N	#1..#3N
OBSOLETE	Sink	Ignored	—	y	y

TABLE 10-1. *Continued*

OH byte	Direction	Process	Location	SDH Regenerator Section RS-0/1/4/16/64/256	SONET Section STS-1/3/12/48/192/768
	Source	Insertion	TT	16 byte string (optinal 0x01H)	1 byte (optional 16 byte string)
J0—section trace identifier	Sink	Defect detection—reporting	TT	y—TIM	n
		PM—N_DS	TT	y	n
		Consequent action—AIS downstream	TT	y	n
		Consequent action—RDI upstream	TT	n/a	n/a
		Protection switching—SF	TT	n/a	n/a
		Acceptance & reporting	TT	y	y
Z0—growth	Source	Insertion—fixed stuff (unscrambled!)	—	(N-1) Z0 bytes	(3N-1) Z0 bytes
	Sink	Ignore	—	y	y
E1—Order Wire	Source	Insertion—external	A	y	y (but not used)
	Sink	Extract	A	y	y (but not used)
F1—user channel	Source	Insertion—internal or external	A	y	n, optional
	Sink	Extract	A	y	n
D1-D3—RS DCC	Source	Insertion	A	y	y
	Sink	Extract	A	y	y
NU—national use	Source	Insertion—fixed stuff or external	—	y	fixed stuff
	Sink	Ignore	—	y	y
	Source	Insertion—fixed stuff: all-zeroes	—	y	y

TABLE 10-2. Multiplex section/Line layer processing

OH byte	Direction	Process	Location	SDH — Multiplex Section MS-0/1/4/16/64/256	SONET — Line STS-1/3/12/48/192/768
B2—BIP24N (defect and PM based on BIP24N violations)	Source	Calculation	TT	y	y
	Sink	Defect detection—reporting	TT	y—dDEG(EB)	y—dEXC/dDEG
		PM—N_EBC	TT	y	y
		Protection Switching—SF [$10e-x$ $x = 3,4,5$]	TT	n on DEG(EB), y on EXC	y [$x = 3$]
		Protection Switching—SD [$10e-x$ $x = 5...9$]	TT	y	y [$x = 6$]
		{TCA} [$10e-x$ $x = 5...9$]	TT	n	y [$x = 6$]
E2—Order Wire	Source	Insertion—external	A	y	y (but not used)
	Sink	Extract	A	y	y (but not used)
D4-D12—MS DCC	Source	Insertion	A	y	y
	Sink	Extract	A	y	y
K1[1-8] K2[1-5(8)] linear APS	Source	Insertion	A	K2[6-8] not used	K2[6-8] used for switching type indication
	Sink	Extraction	A	K2[6-8] not used	K2[6-8] used for switching type indication
K1[1-8] K2[1-8] ring APS	Source	Defect—FOP	A	y	y
	Sink	Insertion	A	y	y
		Extraction	A	y	y
		Defect—FOP	A	y	y
K2[6-8]—RDI [RDI-L]	Source	Insertion	TT	min 1 frame	min 20 frames
	Sink	Defect—RDI	TT	y	y
		PM—F_DS	TT	y	y

TABLE 10-2. *Continued*

OH byte	Direction	Process	Location	SDH Multiplex Section MS-0/1/4/16/64/256	SONET Line STS-1/3/12/48/192/768
K2[6-8]—MS-AIS	Source	—	TT		
	Sink	Defect detection—reporting	TT	y—AIS	y—AIS-L
		PM—N_DS	TT	y	y
		Consequent action—AIS downstream	TT	y	y
		Consequent action—RDI upstream	TT	y	y
		Protection switching—SF	TT	y	y
M1—REI [REI-L] (PM based on BIP24N violations)	Source	Insert BIP violation number	TT	MS-0: n/a MS-1/4/16/64/256: y	STS-1: n/a STS-3/12/48/192/768: y
	Sink	PM—F_EBC	TT	MS-0: n/a MS-1/4/16/64/256: y	STS-1: n/a STS-3/12/48/192/768: y
NU—national use	Source	Insertion—fixed stuff or external	—	y	fixed stuff (all-0's)
	Sink	Ignore	—	y	y
S1—synchronization status message	Source	Insertion	A	SDH code set	SONET code set
	Sink	Extract	A	SDH code set	SONET code set
		Consequent action—by SSM process	A	—	—
[Z1, Z2]—growth	Source	Insertion—fixed stuff: all-zeroes	—	n/a.	y
	Sink	Ignore	—	n/a.	y
Unmarked	Source	Insertion—fixed stuff: all-zeroes	—	y	y
	Sink	Ignore	—	y	y

OH byte	Direction	Process	Location	SDH Administrative unit AU-3/4/4-4c/4-16c/4-64c	SONET STS-1/3c/12c/48c/192c
H1H2—AU {STS} and TU-3 pointer	Source	Pointer generation	A	ss = 10	ss = 10 (old ss = 00)
		Consequent action—AIS downstream	A	250 µs	125 µs
		Increment action	A	y	y
		Decrement action	A	y	y
		New pointer action	A	y	y
		Non-gapped pointer adjustments	A	y	y
		PM—PJE+, PJE−	A	y	n
		PM—{PPJC-Gen}, {NPJC-Gen}, {PJCS-Gen}, {PJCDiff}	A	n	y
	Sink	ss-bits	A	Ignore	Ignore
		Pointer interpretation	A	y (see Note)	y (see Note)
		Increment detection	A	Majority of D bits	Majority of D bits, or 8 out of 10
		Decrement detection	A	Majority of I bits	Majority of I bits, or 8 out of 10
		NDF enabled detection	A	y	y
		Defect—AIS	A	y	y
		Defect—LOP	A	y	y
		PM—{PPJC-Det}, {NPJC-Det}, {PJCS-Det}	A	n	y
		PM—N_DS	TT		y
		Consequent action—AIS (on defect)	A	250 µs	125 µs

TABLE 10-3. *Continued*

OH byte	Direction	Process	Location	SDH — Administrative unit — AU-3/4/4-4c/4-16c/4-64c	SONET STS-1/3c/12c/48c/192c
		Consequent action—AIS (AIS detected)	A	n	y
		Consequent action—RDI upstream	TT	y	y
		Consequent action—{ERDI} server	TT	n	y
		Protection switching—SF	A	y	y
H1H2 AU and {STS} pointer	Sink	AU {STS} multiplex structure determination	A	y MS-SPring limited to the AU's being part of the protection capacity in a MS SPRING ring that are not deselected (locked out). Other AU's may only change type on provisioning.	y BLSR, STS pipe Automatic mode default, Fixed mode selectable per STS-3 group.
H3—AU and {STS} justification opportunity	Source	Insertion—data during justification action	A	y	y
		Insertion—fixed stuff otherwise	A	y	y
	Sink	Extract—data during justification action	A	y	y
		Ignore—otherwise	A	y	y

Note—SONET pointer processing is taken from the defined by [Telcordia GR-253] where the standard [ANSI T1.105] is unclear, i.e.:
1. Pass all 1's H1/H2 byte downstream immediately without waiting for AIS declaration which occurs after three frames.
2. Declare LOP condition if no three consecutive normal or no three consecutive AIS H1/H2 bytes are received in an eight frame window (e.g., alternating AIS/normal pointer will result in LOP).

Table 7.? A higher-order path layer processing

OH byte	Direction	Process	Location	SDH Path VC-3/4/4-Xc (X = 4,16,64,256)	SONET Path STS-1/3c/12c/48c/192c SPE
B3—BIP8	Source	Calculation	TT	y	y
	Sink	Defect detection—reporting	TT	y—dDEG(EB)	y—DEXC/dDEG
		PM—N_EBC	TT	Count errored blocks	Count BIP violations (count errored blocks if ERDI supported)
		Protection switching—SF [10e-x x = 3,4,5]	NIM	n	y [x = 3]
		Protection switching—SD [10e-x x = 5...9]	TT	y	n/a
		{TCA} [10e-x x = 5...9]	NIM	y	y
			TT	n	y [x = 6]
C2—UNEQ	Source	Insertion—all-0's	C	y	y
	Sink	Defect—UNEQ	TT	y, ETSI requires extra robustness for bursts of errors	y
			NIM	y	n
		PM—N_DS	TT	y	n
			NIM	y	y
		Consequent action—AIS downstream	TT	y	n
		Consequent action—RDI upstream	TT	y	y
		Consequent action—{ERDI} upstream	TT	n	n/a
		Protection switching—SF	TT	y	y
			NIM	y	y

TABLE 10-4. *Continued*

OH byte	Direction	Process	Location	SDH Path VC-3/4/4-Xc (X = 4,16,64,256)	SONET Path STS-1/3c/12c/48c/192c SPE
C2—VC-AIS	Source	Creation	TCM	y	n
	Sink	Defect—AIS	TT	n	n
		PM—N_DS	NIM	y	n
			TT	n	n
			NIM	y	n
		Consequent action—AIS downstream	TT	n	n
		Consequent action—RDI upstream	TT	n	n
		Consequent action—{ERDI} server	TT	n	n
		Protection switching—SF	TT	n	n
			NIM	y	n
C2—payload type	Source	Insertion	A	y	y
	Sink	Defect—PLM	A	y	y
			NIM	n	y
		Consequent action—AIS downstream	A	y	y
		Consequent action—RDI upstream	A	n	n
		Consequent action—{ERDI} payload	A	n	y
		Protection switching—SF	A	y	y
			NIM	n	n
		Acceptance & reporting	A	y	y

	Source/Sink				
C2—{PDI} Payload Defect Indicator	Source	Insertion—number of VT's within STS1 in signal fail	A/TT	n	y
	Sink	Defect—{PDI}	NIM	n	y
		Protection switching—SF	NIM	n	y
		Acceptance & reporting	NIM	n	y
F2—user channel	Source	Insertion—internal or external	A	y	y
	Sink	Ignore or extract	A	y	y
F3—user channel	Source	Undefined—fixed stuff insertion	A	y	n/a
	Sink	Ignored	A	y	n/a
{Z3}—growth	Source	Undefined—fixed stuff insertion	A	n/a	y
	Sink	Ignored	A	n/a	y
G1[5]—RDI {RDI-P}	Source	Insertion	TT	Min 1 frame	Min 20 frames
	Sink	Defect—RDI	TT	5 frames	10 frames
		PM—F_DS	NIM	y	y
G1[1-4]—REI {REI-P}	Source	Insertion of BIP violation number	TT	y	y
	Sink	PM—F_EBC	NIM	y	y
G1[5-7]—{ERDI}	Source	Insertion	TT	Count errored blocks	Count BIP violations (count errored blocks if ERDI supported)
			NIM	n	y
	Sink	Defects—{RDIs}, {RDIc}, {RDIp}	A	n	y
			TT	n	y
			A	n	y
			TT	n	y
			NIM	n	y

TABLE 10-4. *Continued*

OH byte	Direction	Process	Location	SDH Path VC-3/4/4-Xc (X = 4,16,64,256)	SONET Path STS-1/3c/12c/48c/192c SPE
	Source	PM (RDIs, RDIc)—F_DS—[ANSI T1.231] does specify its use for PM although this is not realy needed; the use of RDI will be sufficient. RDI = RDIs + RDIc	TT	n	y
	Sink		NIM	n	y
G1[8]—reserved	Source	Insertion—fixed stuff "0"	—	y	y
	Sink	Ignore	—	y	y
H4[1-8]	Source	Insertion—fixed stuff / VCAT overhead	A	y	y
	Sink	Ignore	A	y	y
H4[1-6]—TU multiframe alignment signal	Source	Insertion—fixed stuff "111111"	A	y	y
	Sink	Ignore	A	y	y
H4[7-8]—TU multiframe alignment signal	Source	Insertion	A	y	y
	Sink	Defect—LOM for TU-11/TU-12/TU-2 [VT]	A	y	y
		Consequent action—AIS for TU-11/TU-12/TU-2 {VT}	A	y	y
		Protection switching—SF for TU-11/TU-12/TU-2 {VT}	A	y	y
J1—path trace identifier	Source	Insertion	TT	16 byte string	64 byte string
	C	Insertion—unequipped (all-0's)	C	y	y

Table (continued from previous page — column headers appear on the preceding page). The top rows describe the **Sink** process of the continued overhead byte.

Parameter	Process	AF	Value 1	Value 2
Defect—TIM	Sink	TT	y	y
		NIM	y	n
Fault cause—UNEQ (all zeroes detection)		TT	y (if enabled)	n (ERDI:y)
PM—N_DS		TT	y (if enabled)	n (ERDI:y)
		NIM	n	y (if enabled)
Consequent action—AIS downstream		TT	y	n
Consequent action—RDI upstream		TT	y	y (if enabled)
Consequent action—ERDI-connectivity		TT		n/a
Protection switching—SF		TT		n
		NIM		y
Acceptance & reporting		TT		y
		NIM		y
K3[1-4]—linear APS (under study)	Source	Insertion	A	y	n/a
	Sink	Extraction	A	y	n/a
K3[5-6]—reserved	Source	Insertion	—	—	n/a
	Sink	Ignore	—	—	n/a
K3[7-8]—16 kbit/s path data link (under study)	Source	Insertion	—	—	n/a
	Sink	Extraction	—	—	n/a
{Z4}	Source	Insertion	A	n/a	Fixed stuff
	Sink	Ignore	A	n/a	y
N1—TCM monitoring	Source	Insertion	TCM	y	n/a
	Sink	Extraction	TCM	y	n/a
{Z5}	Source	Insertion	A	n/a	Fixed stuff
	Sink	Ignore	A	n/a	y

TABLE 10-5. Lower-order path layer processing

OH byte	Direction	Process	Location	SDH Path VC-2/12/11	SONET Path VT-6/2/1.5
J2—path trace identifier	Source	Insertion	TT	16 byte string	64 byte string
		Insertion—unequipped (all-zeroes)	C	y	y
	Sink	Defect—TIM	TT	y	n
		Fault cause—UNEQ (all zeroes detection)	NIM	y	n
		PM—N_DS	TT	y	n
		Consequent action—AIS downstream	NIM	y (if enabled)	n (ERDI: y)
		Consequent action—RDI upstream	TT	y (if enabled)	n (ERDI: y)
		Consequent action—ERDI-connectivity	TT	n	y (if enabled)
		Protection switching—SF	TT	y	n
		Acceptance & reporting	TT	y	Under study
			NIM	y	n
K4[1]—extended signal label	Source	Insertion	A	y	y
	Sink	Extraction	A	y	y
K4[2]—VCAT overhead	Source	Insertion	A	y	y
	Sink	Extraction	A	y	y
K4[1-4]—linear APS (under study)	Source	Insertion	A	y	y
	Sink	Extraction	A	—	n/a
K4[5-7]—Reserved	Source	Insertion—fixed stuff ("000" or "111")	—	y	n/a
	Sink	Ignored	—	—	n/a
K4[8]—2kbit/s path data link (under study)	Source	Insertion	—	—	n/a
	Sink	Extraction	—	—	n/a

			1000 µs	500 µs
	Consequent action—AIS downstream	A		
	Increment action	A	y	y
	Decrement action	A	y	y
	New pointer action	A	y	y
	Non-gapped pointer adjustments	A	y	y
	PM—PJE+, PJE−	A	n	n
	PM—PPJC-Gen, NPJC-Gen, PJCS-Gen, PJCDiff	A	n	y
Sink	Pointer interpretation	A	y (note)	y (note)
	Increment detection	A	Majority of D bits	Majority of D bits, or 8 out of 10
	Decrement detection	A	Majority of I bits	Majority of I bits, or 8 out of 10
	NDF enabled detection	A	y	y
	Defect—AIS	A	y	y
	Defect—LOP	A	y	y
	PM—PPJC-Det, NPJC-Det, PJCS-Det	A	n	y
	PM—N_DS	TT	y	y
		NIM	y	y
	Consequent action—AIS (on defect)	A	1,000 µs	500 µs
	Consequent action—AIS (AIS detected)	A	n	y
	Consequent action—RDI upstream	TT	y	y
	Consequent action—ERDI-server	TT	n	y
	Protection switching—SF	A	y	y

Note—Although pointer processing differences between the SONET and SDH standards may be hard to find, Bellcore does impose additional requirements where the standard is unclear. SONET is supposed to also meet the additional Bellcore requirements, i.e.

1. Pass all 1's H1/H2 byte downstream immediately without waiting for AIS declaration which occurs after 3 frames.

2. Declare LOP condition if no 3 consecutive normal or no 3 consecutive AIS H1/H2 bytes are received in an 8 frame window (e.g. alternating AIS/normal pointer will result in LOP).

3. Use 8 out of 10 criteria instead of majority vote for increment/decrement decision (Objective).

TABLE 10-5. *Continued*

OH byte	Direction	Process	Location	SDH Path VC-2/12/11	SONET Path VT-6/2/1.5
V3—TU {VT} justification opportunity	Source	Insertion—data during justification action	A		y
	Sink	Insertion—fixed stuff otherwise	A	y	y
		Extract—data during justification action	A	y	y
V4—reserved	Source	Ignore—otherwise	A	y	y
	Sink	Insertion—fixed stuff	A	y	y
		Ignore	A	y	y
V5[1-2]—BIP2	Source	Calculation	TT	y	y
	Sink	Defect EXC./DEG or BDEG	TT	dBDEG based on 1 second PM info	dEXC/dDEG based on 10^{-x} algorithm
			NIM		
		PM—N_EBC	TT	Count errored blocks	Count BIP violations (count errored blocks if ERDI supported)
			NIM		
		Protection Switching—SF	NIM	n	y
		Protection Switching—SD	TT	y	n/a
			NIM	y	y
V5[3]—REI {REI-V}	Source	Insertion of BIP violation number	TT	y	y
	Sink	PM—F_EBC	TT	y	y
			NIM	y	y
V5[4]—VC-2/12 {VT-6/ VT2} reserved for future use	Source	Insertion—fixed stuff "0"	—	y	y
	Sink	Ignore	—	y	y

Signal	Source/Sink	Process	Type	Req.	Comment
V5[4]—VC11 {VT1.5} RFI	Source	Insertion	TT	y	
	Sink	Defect—RFI (under study)	TT	y	
V5[5-7]—UNEQ	Source	Insertion	C	y	
	Sink	Defect—UNEQ	TT	y	ETSI specification requires extra robustness for bursts of errors
			NIM	n	
		PM—N_DS	TT	n	
			NIM	y	
		Consequent action—AIS downstream	TT	n	
		Consequent action—RDI upstream	TT	y	
		Consequent action—ERDI-connectivity	TT	n/a	
		Protection switching—SF	TT	y	
			NIM		
V5[5-7]—VC-AIS	Source	Creation	TCM	y	
	Sink	Defect—AIS	TT	n	
			NIM	n	
		PM—N_DS	TT	n	
			NIM	n	
		Consequent action—AIS downstream	TT	n	
		Consequent action—RDI upstream	TT	n	
		Consequent action—ERDI-server	TT	n	
		Protection switching—SF	TT	n	
			NIM	n	

TABLE 10-5. *Continued*

OH byte	Direction	Process	Location	SDH Path VC-2/12/11	SONET Path VT-6/2/1.5
V5[5-7]—payload type if value is "101" use the extended signal label in K4[1]	Source	Insertion	A	y	y
	Sink	Defect—PLM	A	y	y
			NIM	n	n
		Consequent action—AIS downstream	A	y	y
		Consequent action—RDI upstream	A	n	n
		Consequent action—ERDI-payload	A	n	y
		Protection switching—SF	A	y	y
			NIM	n	n
		Acceptance & reporting	A	y	y
			NIM	n	n
V5[8]—RDI {RDI-V}	Source	Insertion	TT	min 1 frame	min 20 frames
	Sink	Defect—RDI	TT	5 frames	10 frames
			NIM		
		PM—F_DS	TT	y	y
			NIM	y	y
Z7[5-7]—{ERDI}	Source	Insertion	A	n/a	y
			TT	n/a	y
	Sink	Defects—RDIs, RDIc, RDIp	A	n/a	y
			TT	n/a	y
			NIM	n/a	y
		PM (RDIs, RDIc)—F_DS	TT	n/a	y
			NIM	n/a	y

ABBREVIATIONS

1DM	One-way Delay Measurement
_A	Adaptation function
AcMSI	Accepted MSI values
AcPT	Accepted PT value
ACT	TCM Activation
ADM	Add/Drop Multiplexers
AI	Adapted Information
AIS	Alarm Indication Signal
ANSI	American National Standards Institute
AP	Access Point
API	Access Point Identifier
APR	Automatic optical Power Reduction
APS	Automatic Protection Switching
ATM	Asynchronous Transfer Mode
AU–n	Administrative Unit (n = 3, 4)
AUG-N	Administrative Unit Group of level N (N = 1, 4, 16, 64, 256)
B3ZS	Bipolar with Three-Zero Substitution
BDI	Backward Defect Indication

The ComSoc Guide to Next Generation Optical Transport: SDH/SONET/OTN,
by Huub van Helvoort
Copyright © 2009 Institute of Electrical and Electronics Engineers

BEI	Backward Error Indication
BIAE	Backward Incoming Alignment Error
BIP	Bit Interleaved parity
BIP-N	Bit Interleaved Parity N-bits
BLSR	Bi-directional Line Switched Ring
_C	Connection function
CBR	Constant Bit-Rate
CCAT	Contiguous conCATenation
CCC	Continuity and Connectivity Check
cHEC	core Header Error Check
CI	Characteristic Information
CLP	Cell Loss Priority
CMI	Coded Mark Inversion
CP	Connection Point
CRC	Cyclic Redundancy Check
CTRL	Control
DA	Destination Address
DCC	Data Communications Channel
DCCM	Multiplex Section DCC
DCCMx	Extended MS DCC
DCCR	Regenerator Section Data Communications Channel
DEG	Degraded signal
DEG	DEGraded; based on 10^{-x} algorithm
DMM DMR	Delay Measurement; ITU-TDM, Message and Response
DNU	Do Not Use
DQDB	Distributed Queue Dual Bus
DSAP	Destination Service Access Point
DXC	Digital Cross Connect
EC-N	Electrical Carrier Level N (N = 1, 3)
ECSA	Exchange Carriers Standards Association
EDC	Error Detection Code
eHEC	extension HEC
EMF	Element Management Function
EOS	End of Sequence
EOW	Engineering Order Wire
{ERDI}	Enhanced RDI
ES1	Electrical Section level 1
ETH	Ethernet MAC layer
ETSI	European Telecommunications Standards Institute

ETY	Ethernet Physical layer
EXC	Excessive error
EXC	EXCessive based on 10^{-x} algorithm
EXP	EXPerimental
ExTI	Expected Trail Identifier
F_DS	Far end—Defect Second
F_EBC	Far end—Errored Block Count
FAS	Frame Alignment Signal
FC-BBW	Fibre Channel BackBone WAN
FCS	Frame Check Sequence
FDDI	Fibre Distributed Data Interface
FDI	Forward Defect Indication
FDM	Frequency Division Multiplex
FEC	Forward Error Correction
FOP	Failure Of Protocol
FSI	FEC Status Indicator
FTFL	Fault Type and Fault Localization communication channel
GCC	Generic Communications Channel
GFC	Generic Flow Control
GFP	Generic Framing Procedure
GFP-F	Frame mapped GFP
GFP-T	Transparent mapped GFP
GID	Group IDentity
GMP	Generic Mapping Procedure
HDLC	High-Level Data Link Control
HEC	Header Error Control
IAE	Incoming Alignment Error
IEC	Incoming Error Count
ITU-T	International Telecommunication Union— Telecommunication standardization sector (was CCITT)
JC	Justification Control
JOH	Justification OverHead
LAN	Local Area Networks
LBM LBR	LoopBack, Message and Response
LC	Link connection
LCAS	Link Capacity Adjustment Scheme
LCC	Logical Link Control
LCK	Lock
LMM LMR	Loss Measurement, Message and Response

LOC	Loss Of Continuity
LOF	Loss Of Frame
LOM	Loss Of Multiframe
LOP	Loss Of Pointer
LOS	Loss Of Signal
LSB	Least Significant Bits
LSS	Loss of Sequence Synchronization
LTC	Loss of Tandem Connection
LTM LTR	Link Trace, Message and Response
MAN	Metro Area Networks
MCC	Management Communications Channel
ME	Maintenance Entities
MEL	Maintenance Entity Level
MEP	Maintenance End Points
MFAS	Multi-Frame Alignment Signal
MFI	Multi-Frame Indicator
MIP	Maintenance Intermediate Points
MMG	MisMerGe
MP	Management Point
MPLS	Multiprotocol Label Switching
MPLS-TP	MPLS Transport Profile
MSAP	Multi-Service Access Platforms
MSB	Most Significant Bits
MSI	Multiplex Structure Identifier
MSIM	MSI Mismatch
MSn	Multiplex Section level n (n = 1, 4, 16, 64, 256)
MSOH	Multiplex Section OverHead
MSPP	Multi-Service Provisioning Platforms
MSSP	Multi-Service Switching Platforms
MST	Member STatus
MSTP	*Multi-Service Transport Platforms*
MTBF	Mean Time Between Failures
MTP	MPLS-TP Path
MTS	MPLS-TP Section
MTTR	Mean Time To Repair
N_DS	Near end—Defect Second
N_EBC	Near end—Errored Block Count
NC	Network connections
NDF	New Data Flag

NGN	Next Generation Network
NIM	Non-Intrusive Monitors
NJO	Negative Justification Opportunity
NNI	Network–Node Interface
{NPJC}	Negative Pointer Justification Counter
NUT	Non-preemptible Unprotected Traffic
OCh	Optical Channel layer
OC-N	Optical Carrier of level N (N = 1, 3, 12, 48, 192, 768)
ODI	Outgoing Defect Indication
ODUk	Optical channel Data Unit of level k (k = 1, 2, 3)
OEI	Outgoing Error Indication
{OFS}	Out of Frame Second
OH	OverHead
OMS	Optical Multiplex Section
OOS	OTM Overhead Signal
OPS	Optical Physical Section
OPUk	Optical channel Payload Unit of level k (k = 1, 2, 3)
OSC	Optical Supervisory Channel
OSn	Optical Section of level n (n = 1, 4, 16, 64, 256)
OTM	Optical Transport Module
OTN	Optical Transport Network
OTUk	Optical channel Transport Unit of level k (k = 1, 2, 3)
OUI	Organizationally Unique Identifier
PCC	Protection Communication Channel
PCM	Pulse Coded Modulation
PDH	Plesiochronous Digital Hierarchy
{PDI}	Payload Defect Indicator
PDU	Protocol Data Unit
PID	Protocol Identification
{PJCDiff}	Pointer Justification Count Difference
{PJCS}	Pointer Justification Count Seconds
PJE+	positive Pointer Justification Event
PJE–	negative Pointer Justification Event
PJO	Positive Justification Opportunity
PLM	PayLoad Mismatch
PM	Performance Monitoring
PMI	Payload Missing Indication
POH	Path Overhead
PP	Replication Point

{PPJC}	Positive Pointer Justification Counter
PPP	Point to Point Protocol
PRBS	Pseudo Random Bit Sequence
PSI	Payload Structure Identifier
PSL	Payload Signal Label
PT	Payload Type
PTC	Packet Transport Channel
PTD	PTN diagnostic function
PTI	Payload Type Indicator
PTN	Packet Transport Network
PTP	Packet Transport Path
PTS	Packet Transport Section
QoS	Quality of Service
RDI	Remote Defect Indication
REI	Remote Error Indication
RES	Reserved
RFI	Remote Failure Indication
RP	Remote Point
RS-Ack	Re-Sequence Acknowledge
RSn	Regenerator Section of level (n = 1, 4, 16, 64, 256)
RSOH	Regenerator Section OverHead
SA	Source Address
SAN	Storage Area Networks
SAP	Service Access Point
SAPI	Regenerator Section Access Point Identifier; ITU-T
SD	Signal Degrade
SDH	Synchronous Digital Hierarchy
{SEFS}	Severely Errored Frame Second
SF	Signal Fail
_Sk	Sink function
SM	Section Monitoring
SNAP	Sub Network Access Protocol
SNC	Sub-Network Connection
SNC/I	SNC—Inherent monitored
SNC/N	SNC—Non-intrusive monitored
SNC/S	SNC—Sub-layer trail monitored
SNC/T	SNC—Test monitored
SNCP	SNC Protection
_So	Source function

SoF	Start of Frame
SONET	Synchronous Optical NETwork
{SPE}	Synchronous Payload Eveloppe
SPRing	Shared Protection Ring
SQ	SeQuence number
SSAP	Source Service Access Point
SSD	Server Signal Degrade
SSF	Server Signal Fail
SSM	Synchronization Status Message
STM-N	Synchronous Transport Module of level N (N = 1, 4, 16, 64, 256)
{STS-N}	Synchronous Transport Signal of level N (N = 1, 3, 12, 48, 192, 768)
TC_OH	TCM OverHead
{TCA}	Threshold Crossing Alert
TCM	Tandem Connection Monitoring
TCP	Termination CP
TDM	Time Division Multiplex
tHEC	type HEC
TIM	Trace Identifier Mismatch
TM	Terminating Multiplexers
TP	Timing Point
TSD	Trail Signal Degrade
TSE	Test Sequence Errors
TSF	Trail Signal Fail
TST	Test
_TT	Trail Termination function
TTI	Trail Trace Identifier
TTL	Time To Live
TU–n	Tributary Unit of type n (n = 2, 12, 11)
TUG–n	TU Group of type n (n = 2, 12, 11)
UML	Unexpected MEL
UMP	Unexpected MEP
UNEQ	Unequipped
UNI	User–Network Interface
UPI	User Payload Identifier
UPSR	Unidirectional Path Switched Ring
USR	User Channel
VC–n	Virtual Container of type n (n = 4, 3, 2, 12, 11)

VC–n–Xc	Contiguous concatenated VC-n
VC–n–Xv	Virtual concatenated VC-n
VCAT	Virtual conCATenation
VCG	Virtual Concatenation Group
VCI	Virtual Channel Identifier
VP	ATM Virtual Path
VPI	VP Identifier
VPN	Virtual Private Networks
VSP	Vendor SPecific
{VT}	Virtual Tributary
WAN	Wide Are Networks

REFERENCES

[ANSI NCITS 342]	ANSI NCITS 342: *Fibre Channel—Backbone (FC-BB)*
[ANSI T1.102]	ANSI T1.102: *Digital Hierarchy—Electrical Interfaces*
[ANSI T1.105]	ANSI T1.105: *Synchronous Optical Network (SONET)— Basic Description including Multiplex Structure, Rates and Formats*
[ETSI EN 300 417-1...10]	ETSI EN 300 417: *Transmission and Multiplexing (TM); Generic requirements of transport functionality of equipment; Part 1-1: Generic processes and performance Part 2-1: "Synchronous Digital Hierarchy (SDH) and Plesiochronous Digital Hierarchy (PDH) physical section layer functions"; Part 3-1: "Synchronous Transport Module-N (STM-N) regenerator and multiplex section layer functions"; Part 4-1: "Synchronous Digital Hierarchy (SDH) path layer functions"; Part 5-1: "Plesiochronous Digital Hierarchy (PDH) path layer functions"; Part 6-1: "Synchronization layer functions"; Part 7-1: "Equipment management and auxiliary layer functions"; Part 9-1: "Synchronous Digital Hierarchy (SDH) concatenated path layer functions; Requirements". Part 10-1: Synchronous Digital Hierarchy (SDH) radio specific functionalities*
[ETSI ETS 300 216]	ETSI ETS 300 216: *Network Aspects (NA); Metropolitan Area Network (MAN); Physical layer convergence procedure for 155,520 Mbit/s*

The ComSoc Guide to Next Generation Optical Transport: SDH/SONET/OTN, by van Helvoort
Copyright © 2009 Institute of Electrical and Electronics Engineers.

[IEEE 802.3ae]	IEEE 802.3ae: *IEEE Standard for Information technology—Telecommunications and information exchange between systems—Local and metropolitan area networks—Specific requirements—Part 3: CSMA/CD Access Method and Physical Layer Specifications. Amendment: MAC Parameters, Physical Layers, and Management Parameters for 10 Gbit/s Operation*
[IETF RFC1661]	IETF RFC1661: *The Point-to-Point Protocol (PPP)*
[IETF RFC1662]	IETF RFC1662: *PPP in HDLC-like Framing*
[IETF RFC3032]	IETF RFC3032: *MPLS Label Stack Encoding*
[ITU-T Rec. G.664]	ITU-T Recommendation G.664: *Optical safety procedures and requirements for optical transport systems*
[ITU-T Rec. G.691]	ITU-T Recommendation G.691: *Optical interfaces for single channel STM-64 and other SDH systems with optical amplifiers*
[ITU-T Rec. G.703]	ITU-T Recommendation G.703: *Physical/electrical characteristics of hierarchical digital interfaces*
[ITU-T Rec. G.704]	ITU-T Recommendation G.704: *Synchronous frame structures used at 1,544, 6,312, 2,048, 8,448 and 44,736 kbit/s hierarchical levels*
[ITU-T Rec. G.707]	ITU-T Recommendation G.707/Y.1322: *Network node interface for the synchronous digital hierarchy (SDH)*
[ITU-T Rec. G.709]	ITU-T Recommendation G.709/Y.1331: *Interfaces for the Optical Transport Network (OTN)*
[ITU-T Rec. G.743]	ITU-T Recommendation G.743: *Second order digital multiplex equipment operating at 6,312 kbit/s and using positive justification*
[ITU-T Rec. G.751]	ITU-T Recommendation G.751: *Digital multiplex equipments operating at the third order bit rate of 34,368 kbit/s and the fourth order bit rate of 139,264 kbit/s and using positive justification*
[ITU-T Rec. G.752]	ITU-T Recommendation G.752: *Characteristics of digital multiplex equipments based on a second order bit rate of 6,312 kbit/s and using positive justification*
[ITU-T Rec. G.753]	ITU-T Recommendation G.753: *Third order digital multiplex equipment operating at 34,368 kbit/s and using positive/zero/negative justification*
[ITU-T Rec. G.754]	ITU-T Recommendation G.754: *Fourth order digital multiplex equipment operating at 139,264 kbit/s and using positive/zero/negative justification*
[ITU-T Rec. G.783]	ITU-T Recommendation G.783: *Characteristics of synchronous digital hierarchy (SDH) equipment functional blocks*
[ITU-T Rec. G.798]	ITU-T Recommendation G.798: *Characteristics of optical transport network hierarchy equipment functional blocks*
[ITU-T Rec. G.7041]	ITU-T Recommendation G.7041/Y.1303: *Generic framing procedure (GFP)*

[ITU-T Rec. G.7042]   ITU-T   Recommendation   G.7042/Y.1305:   *Link   capacity adjustment scheme (LCAS) for virtual concatenated signals*

[ITU-T Rec. G.805]   ITU-T Recommendation G.805: *Generic functional architecture of transport networks*

[ITU-T Rec. G.806]   ITU-T   Recommendation   G.806:   *Characteristics   of transport   equipment—Description   methodology   and   generic functionality*

[ITU-T Rec. G.808.1]   ITU-T Recommendation G.808.1: *Generic protection switching—Linear trail and subnetwork protection*

[ITU-T Rec. G.809]   ITU-T   Recommendation   G.809:   *Functional   architecture   of connectionless layer networks*

[ITU-T Rec. G.811]   ITU-T   Recommendation   G.811:   *Timing   characteristics   of primary reference clocks*

[ITU-T Rec. G.812]   ITU-T Recommendation G.812: *Timing requirements of slave clocks suitable for use as node clocks in synchronization networks*

[ITU-T Rec. G.813]   ITU-T   Recommendation   G.813:   *Timing   characteristics   of SDH equipment slave clocks (SEC)*

[ITU-T Rec. G.825]   ITU-T   Recommendation   G.825:   *The   control   of   jitter   and wander within digital networks which are based on the synchronous digital hierarchy (SDH)*

[ITU-T Rec. G.841]   ITU-T Recommendation G.841: *Types and characteristics of SDH network protection architectures*

[ITU-T Rec. G.8021]   ITU-T   Recommendation   G.8021/Y.1341:   *Characteristics   of Ethernet transport network equipment functional blocks*

[ITU-T Rec. G.8031]   ITU-T Recommendation G.8031/Y.1342: *Ethernet linear protection switching*

[ITU-T Rec. G.8032]   ITU-T Recommendation G.8032/Y.1344: *Ethernet ring protection switching*

[ITU-T Rec. G.8121]   ITU-T   Recommendation   G.8121/Y.1381:   *Characteristics   of MPLS-TP equipment functional blocks*

[ITU-T Rec. G.8131]   ITU-T   Recommendation   G.8131/Y.1382:   *Linear   protection   switching   for   MPLS   Transport   Profile   (MPLS-TP) networks*

[ITU-T Rec. G.8132]   ITU-T Recommendation G.8132: *MPLS-TP Shared Protection Ring (MT-SPRing)*

[ITU-T Rec. G.957]   ITU-T Recommendation G.957: *Optical interfaces for equipments   and   systems   relating   to   the   synchronous   digital hierarchy*

[ITU-T Rec. G.959.1]   ITU-T Recommendation G.959.1: *Optical transport network physical layer interfaces*

[ITU-T Rec. I.150]   ITU-T Recommendation I.150: *B-ISDN asynchronous transfer mode functional characteristics*

[ITU-T Rec. I.361]   ITU-T   Recommendation   I.361:   *B-ISDN   ATM   layer specification*

[ITU-T Rec. I.432.1]   ITU-T Recommendation I.431.1: *B-ISDN user-network interface—Physical layer specification: General characteristics*

[ITU-T Rec. X.200]   ITU-T Recommendation X.200: *Information technology—Open Systems Interconnection—Basic Reference Model: The basic model*

BIBLIOGRAPHY

IEEE Communications, March 1989, "SONET: Now It's the Standard Optical Network."

Next Generation SDH/SONET: Evolution or Revolution, John Wiley & Sons, ISBN 0-470-09120-7.

SDH/SONET Explained in Functional Models: Modeling the Optical Transport Network, John Wiley & Sons, ISBN 0-470-09123-1.

The ComSoc Guide to Next Generation Optical Transport: SDH/SONET/OTN,
by Huub van Helvoort
Copyright © 2009 Institute of Electrical and Electronics Engineers

INDEX

The ComSoc Guide to Next Generation Optical Transport: SDH/SONET/OTN,
by Huub van Helvoort
Copyright © 2009 Institute of Electrical and Electronics Engineers

Printed and bound by CPI Group (UK) Ltd, Croydon, CR0 4YY

27/10/2024

14580257-0002